Cave and Karst Systems of the World

Series Editor

James W. LaMoreaux, P.E.LaMoreaux and Associates, Tuscaloosa, AL, USA

More information about this series at http://www.springer.com/series/11987

Ali Yamaç • Eric Gilli •
Ezgi Tok • Koray Törk

Caves and Karst of Turkey—Vol. 1

History, Archaeology and Caves

Springer

Ali Yamaç
OBRUK Cave Research Group
Istanbul, Turkey

Ezgi Tok
Eurasia Institute of Earth Sciences
Istanbul Technical University
Istanbul, Turkey

Eric Gilli
Department of Geography
University Paris 8
Paris, France

Koray Törk
Geological Research
Mineral Research and Exploration of Turkey
Ankara, Turkey

ISSN 2364-4591 ISSN 2364-4605 (electronic)
Cave and Karst Systems of the World
ISBN 978-3-030-65500-6 ISBN 978-3-030-65501-3 (eBook)
https://doi.org/10.1007/978-3-030-65501-3

This Springer imprint is published by the registered company Springer Nature Switzerland AG
The registered company address is: Gewerbestrasse 11, 6330 Cham, Switzerland

Preface

For the first—almost—three decades of Turkish caving, there were only two speleology organizations in Turkey: Cave Research Association (MAD) established by Dr. Temuçin Aygen in 1964 and Boğaziçi University Speleological Society (BÜMAK) established in 1973.

On the other hand, during those dates, speleology was accepted as both a sport and a discipline all around the world for a long time. Apart from the works and studies carried out in other countries, even in Turkey, there were numerous speleological and biospeleological researches carried out by foreigners years ago or ongoing on those dates. Various foreign explorers and academicians had already published numerous articles and even books about the caves of Turkey.

After all these years, the number of speleological associations in Turkey has reached 18 today and numerous different associations, including the governmental organizations, are actively caving. In the preface of proceedings of the 1. National Speleology Symposium in 1990, Prof. Mustafa Aktar had stated that "Speleology, by extending to numerous areas such as sport, art, and tourism during its development process which started with science has achieved a cultural structure. And in Turkey, the number of people involved in this area as a profession or as an amateur interest has increased and organizations have emerged."

Indeed, with the efforts of all these new associations and organizations, over the years, hundreds of new caves have been explored and researched. Not only with the newly explored caves, but also with the new branches of the previously explored caves, the lengths and depths of the caves constantly change each year. Even during the writing phase of this book, we have witnessed the change of the list of the longest and deepest caves.

Independent from all these explorations performed, numerous different scientific studies are being carried out in recent years related to the caves of Turkey. The works and studies in the areas of cave biology and paleoclimatology, which have gained a huge acceleration especially in recent years, are a sign of the bright and promising future of the scientific cave studies in Turkey. Various researches are being carried out in recent years regarding these issues, and numerous articles have been published. Likewise, in the area of cave archaeology, rather than old-style rough researches, precise scientific works and studies are being carried out. Due to the rapidly developing new techniques, not only in the archaeological excavations but also in the dating and DNA tests, striking results are being achieved.

This book explains the geology, hydrogeology, and exploration stories of 45 caves of Turkey, chosen among the thousands of caves explored and researched until now. Additionally, the findings of 20 caves chosen among 90 archaeological caves researched or excavated until now are explained in a separate chapter.

However, as explained comprehensively in Chap. 4 of this book, the karstic area of Turkey is larger than the total area of numerous countries and for a country with such a large karstic area, neither the number of the speleological associations existing today, nor the number of explorations and scientific works carried out is sufficient. This country, which has tens of thousands of unexplored caves, requires a significant acceleration in the areas of both scientific and sportive cave explorations and cave archaeology and more importantly needs academic organizations to carry out karst and cave works and studies.

We wish that this book enlightens the immensity of the work required to be carried out and hope that it will form a basis for the new researches.

Istanbul, Turkey

Ali Yamaç
Eric Gilli
Ezgi Tok
Koray Törk

Acknowledgements During the preparation of this book, we had used hundreds of references but, without the colossal work of Bülent Erdem, "Bibliography of Speleology in Turkey" and his dedicated efforts, it would be impossible to be informed and to reach all these sources. We would like to express our sincere gratitude to him for his enormous help.

Contents

About the Authors

Ali Yamaç is a Turkish speleologist. He was born in 1957 and caving since 1976. He had explored and surveyed hundreds of caves during those years, of which some are among the deepest and longest caves of Turkey. He was President of Speleological Federation of Turkey during 2007–2009 and President during the establishment of three different caving organizations. He was the leader of the team who had prepared the Cave Inventory of Turkey. In addition to his natural cave explorations, as a part of his artificial cave projects, he had explored and surveyed numerous underground cities and rock dwelled settlements around different regions of Turkey.

Eric Gilli is a French karstologist. He was born in 1957 and studied geology and physical geography in Nice, Paris, and Aix-Marseille. Formerly a consultant in Nice, he has been since 2001 Professor at the Geography Department at the University of Paris 8. His main topics are hydrogeology, karst, caves, groundwater, palaeo-environments, earthquakes, tsunamis, and natural disasters. He has also been involved in shipwreck explorations and troglodyte studies. His international studies have taken him to many parts of world.

Ezgi Tok is a Turkish speleologist. Her research is focused on microbial life in cave environments. She explored several caves all over Turkey to reveal the microbial diversity, the ecosystem functioning of microbial populations, and their relation with the psychochemical conditions within the cave ecosystem. She had been caving since 2007 and joined several national and international speleological expeditions in Turkey and foreign. She also had surveyed and mapped several caves of Turkey; among them İnsuyu Cave, which is the longest, and Kocain Cave, which is the largest chamber of Turkey, are especially important.

Koray Törk is a speleologist and karst hydrogeologist working in most of the karst areas of Turkey. He jointed to Cave Research Association (MAD) as a caver and then have been Member of Mineral Research and Exploration of Turkey (MTA) as a speleologist. He has been working predominantly on karst base natural hazards (sinkhole collapse/obruk) and cave protection for more than ten years.

Abbreviations

AKUMAK	Akdeniz University Caving Club
ASPEG	Anatolian Speleology Group
BSU	Balkan Speleological Union
BUMAD	Boğaziçi International Speleology Association
BÜMAK	Boğaziçi University Speleological Society
CMN	Club Martel de Nice
FSE	European Speleological Federation
ITUMAK	Istanbul Technical University Speleology Club
LUSS	Leicester University Speleological Society
MAD	Cave Research Association
MTA	General Directorate of Mineral Research and Exploration
OBRUK	Obruk Cave Research Group
SCP	Speleo Club de Paris
TMF	Speleological Federation of Turkey
UIS	International Union of Speleology

Speleology in Turkey

1

Abstract

Excluding individual studies of some researchers, speleological researches in Turkey started very late, in the 1960s, and they were largely conducted by foreign cavers until a very recent date, the 1980s. In this period and after, numerous caves were found in the Taurus Mountains through the explorations by foreign cavers. The current state of Turkish caving that accelerated only after the 1980s is not enough for a country with such a large karstic area. There are only 18 caving organizations in Turkey, and they generally work without a long-term plan and can not conduct full regional surveys. Despite these negative facts, about 3000 caves were researched and surveyed in the past 40–50 years.

1.1 Precursors

The first known cave research in Turkey is conducted by Dr. Abdullah Bey in Yarımburgaz Cave, near İstanbul. He was born in Vienna in 1801 as Karl Eduard Hammerschmidt. As a multidisciplinary intellectual, Hammerschmidt was an Austrian geologist, entomologist, and physician. After getting his law degree from the University of Vienna in 1827, he studied medicine with an emphasis on anesthesiology. He became a medical doctor in 1837 and worked both as a zoology teacher and a surgeon at the University of Vienna (Fig. 1.1).

He joined the uprising also known as the Vienna Revolution in 1848. After the suppression of the uprising, he entered Turkey along with other revolutionaries from Hungary and was employed as a teacher of medicine, zoology, and mineralogy in the medical school of Istanbul. When Austria demanded Turkey to deport him, he was transferred to Damascus where he worked as a hospital physician. In this term of his life, Hammerschmidt converted to Islam and took on the name "Abdullah Bey." He served in the Turkish army during the Crimean War.

In 1862, he joined the faculty of medicine in Istanbul, teaching geology, mineralogy, and zoology. He founded the Natural History Museum of the Imperial Medical School of Constantinople, with many fossils he gathered around Istanbul over many years. One of the new species found among these fossils was given the name "*Cryphaeus abdullahi.*" He was among the founders of the Turkish Red Crescent. He died in 1874 in Istanbul. In his short article, "Die Umgebung des See's Kütchücktschekmetché in Rumelien" which was published in the 12. issue of "Verhandlungen der k.k. Geologischen Reichsanstalt," Abdullah Bey explains the examinations he performed in Yarımburgaz Cave near Istanbul. In his article, Abdullah Bey, who found the Yarımburgaz Cave while performing geological research in the area, explains the chambers, the steps, the structure resembling an altar stone, and the arched cells carved in the rocks found at the cave entrance. He roughly gives the measurements of this area and writes about thinking of this place to be a living quarter or a secret meeting place for people in ages past. Walking a few hundred steps in the main gallery, Abdullah Bey explains the speleothems he encounters in detail (Bey, 1869).

Three years after the publication of this first article of his, Abdullah Bey's second article named "Études Géologiques sur les Environs de Constantinople. Yarym-Bourgas, Macri-Keuy, Sary-Keuy" was published in Gazette Médicale d'Orient. This second article is quite different from the previous one and almost completely focuses on geology. He writes about having visited the Yarımburgaz Cave again during the three years between his articles and notes having traveled 250 m into the cave taking measurements and examining rock formations, even though he could not have reached the end of the cave. He writes that the area was composed of Miocene old limestone and he lists and names the fossils he finds in the area. He explains some of these fossils he

A. Yamaç et al., *Caves and Karst of Turkey—Vol. 1*, Cave and Karst Systems of the World,
https://doi.org/10.1007/978-3-030-65501-3_1

Fig. 1.1 Karl Hammerschmidt, lithograph by Josef Kriehuber, 1836

encounters like *Congeria subglobosa*, *Congeria cordiformis*, and *Pectunculus fichteli* more in detail (Bey, 1872).

Rabius Bousquet's article "Les Grottes de Yarim-Bourgas," which was published in the Echos d'Orient journal in 1901, is an unexpectedly incompetent expedition report considering its name. In the two-page article, he tells about points of interest between Istanbul and the Yarımburgaz Cave, gives a general description of the upper entrance of the cave but almost completely skips progressing in the main gallery (Bousquet, 1901). The French caver E.A. Martel came to Turkey in 1910 and visited few caves near Konya, Sille, and Ankara (Martel, 1911).

In his Turkish article called "How to Survey a Cave and a Visit to Yarımburgaz Cave" published in 1918 in the journal Tedrisat Mecmuası, Harun Reşit Kocacan explains in detail how a cave should be examined scientifically. Although he reached the end of the Yarımburgaz Cave, a map for the cave could not have been drawn (Kocacan, 1918).

While different researchers were trying to examine the Yarımburgaz Cave, two archaeologists started much more serious and scientific cave research in southern Anatolia, Antalya. As one of the losing parties of World War I, the Ottoman Empire signed the Armistice of Mudros in 1918. Parallel to this agreement, the Italian army occupied southern Anatolia and Antalya in 1919. Italian archaeologists who arrived in the area with the army had explored and researched many ancient settlements until the end of the occupation in 1921. Being different from archaeologist Vittorio Viale's archaeological excavation in Gurma Cave (Viale, 1925–26), Giuseppe Moretti's research in Kocain Cave can be considered both an extensive cave exploration and a survey. Moretti, who conducted an ancient city research north of Antalya around Camili Village, was told that there were no ruins in the vicinity when he reached the area. On the other hand, Ahmet Aga, who lived alone in a valley northeast of Camili Village as a woodcutter, mentioned Moretti about a huge cave found in the mountains with writings in it. Spending the night in the area over this information, Moretti and his team reached Kocain on July 30, 1919. After a rough examination, Moretti understood that this research would take long and returned, to come back to Kocain again on September 5, 1919 (Fig. 1.2). Professor Azeglio Berretti from the Museum of Rome also attended this second expedition. Having measured the cave, Moretti writes in his article "In Daghindà Quogia In, La Grande Caverna nella Montagna delle Caverne" which was published in "Annuario del la Reale Scuola Archeologica di Atene" in 1924, that the cave entrance is 60 meters wide and 15 meters high. He specifies that the width of the cave changes between 50 and 60 m and height was around 25 m throughout the gallery. As the measurements taken from the hall start to get wider at the end of the gallery, the error rate increases a little bit more. Not being able to see the ceiling from that point on, Moretti could not comment on the height and only writes that it has "very unusual dimensions." He measures the diameter of the column, which resides behind the hill 150 meters farther from the point where the cave rotates toward the south, to be 12–13 m. Moretti draws the cistern and the inscriptions found inside the cave and also takes many pictures. The Kocain map published in his article is the very first cave map drawn in Anatolia. Even though there are differences between Moretti's map and other

Fig. 1.2 Kocain Cave from the first chamber toward the entrance. Dated 1919, this is possibly one of the earliest photographs of a cave in Turkey (Moretti, 1923–1924)

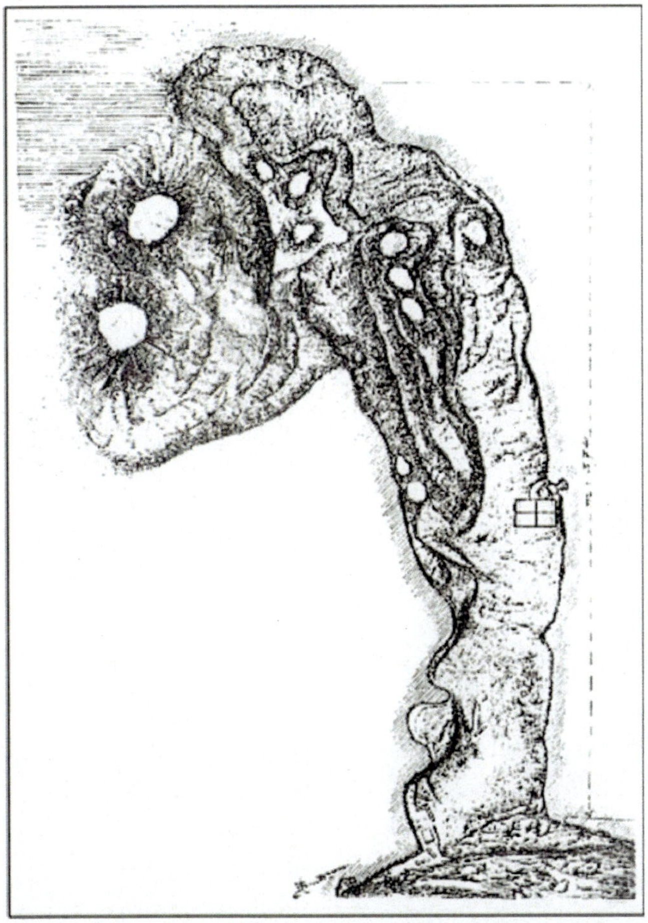

Fig. 1.3 Map of Kocain Cave (Moretti, 1923–1924)

Fig. 1.4 Map of Yarımburgaz Cave (Hovasse, 1927)

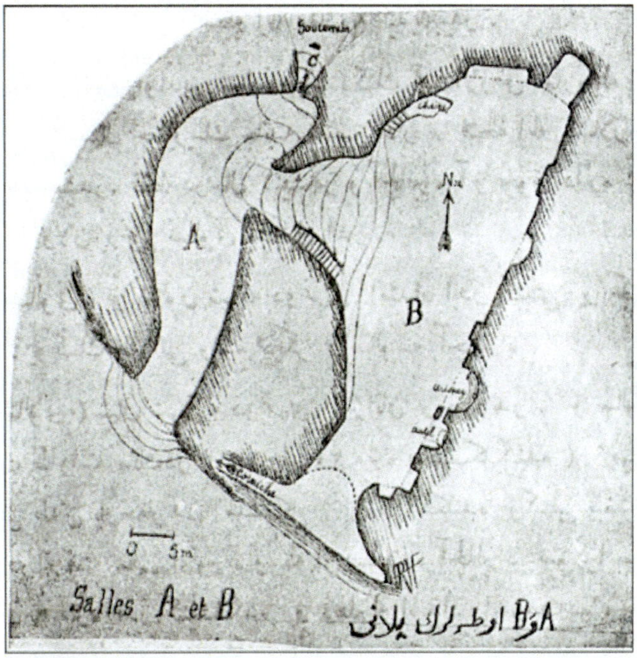

Fig. 1.5 Detailed plan of the first two chambers of Yarımburgaz Cave showing the apsis of the church and niches carved from the main rock (Hovasse, 1927)

Kocain maps drawn later, it is admirable in terms of the applied technique and delicacy (Fig. 1.3) (Moretti, 1923–1924).

Following Moretti's Kocain Cave map, ongoing research since 1869 on Yarımburgaz Cave also reaches a happy conclusion. Professor Raymond Hovasse publishes an extensive article about this cave and draws a map of it in 1927 (Fig. 1.4). Born in France in 1895, Hovasse concluded his education in 1919 which was disrupted by World War I and started to work in Sorbonne University Histology Laboratory the following year. With an invitation from foreign affairs ministry, he started giving zoology lectures in Istanbul University Science Faculty and continued his lectures until leaving the country in 1932. During his time in Istanbul, Hovasse established the Baltalimanı Zoology Station and published nine different articles (Kadioğlu, 2003). His first published article "Yarımburgaz Mağarası (La Grotte de Yarım Bourgas)" was published in Istanbul University Science Faculty Journal in 1927. In this systematical research he conducted with Prof. René Jeannel, he measured, drew, and took pictures of every formation carved inside the rocks found within the upper cave (Fig. 1.5). He made a thorough geological and hydrogeological examination of the cave and realized that the underground creek flowing through the cave went further down in time and carved the Yarımburgaz Cave, shaping it into a vadose state. The most important part of this 26-page article is about the life forms inside the cave, which is Jeannel's area of expertise. He explains 17 different life forms he encountered, one by one and in detail. Among these, there are two different types of bats. Other than these two different types he identified as Rhinolophus Ferrumequinum and Miniopterus Schraibersi Mapfre, he found 15 different types of insects. Among these, especially two types of troglobites, Glomeridae and Collembola stand out the most (Hovasse, 1927) .

As far as we know, after this thorough study of Hovasse, there had not been such extensive speleological research in Turkey for many years. After the aforementioned Yarımburgaz Cave map, the third cave map to be drawn in Turkey is Hans Henning von der Osten's simple sketch of a cave explored near Erzincan–Kemaliye in 1927 (Von der Osten, 1929).

All the speleological explorations conducted in Turkey after the 1920s play second string to "cave archaeology" mentioned in Chap. 2 and "biospeleology" mentioned in Chap. 3. Archaeologists conducting researches in caves had done some cave surveys in this period, but these studies are far from including any geological and geomorphological data. The same problem is also encountered in biospeleological researches. Even though Şevket Aziz Kansu's assistant archaeologist Kılıç Kökten examined and drew maps during the İnönü Caves archaeological excavations that began in 1938, these studies could not go further than being a tool for these excavations (Kansu, 1939).

In a contemporary sense, speleological studies in Turkey began with Temuçin Aygen (1921–2003). Aygen is considered to be the founder of Turkish speleology in both a scientific and a sportive sense, as he transformed cave sports and studies to be understood as teamwork (Fig. 1.6). MAD, which he founded and became the founding president of, is the first cave association in Turkey (Altay, 2007). Aygen traveled around Anatolia for almost 50 years and displayed an amazing work in terms of registering and protecting hundreds of caves he explored during that time. Aside from the fact that he explored many of the most important and well-known caves in Turkey, many caves started being protected and five different areas have "national park" status thanks to him (Aygen, 1988).

As a graduate of the University of Geneva, Faculty of Geology, Aygen got his doctorate from the University of Istanbul Faculty of Geology in 1951 and started working as a geologist in MTA the same year. From 1952 and onward, he conducted researches in various regions of Turkey on water distribution of karstic areas and the productivity of hydroelectricity plants. He started taking an interest in caves during these researches. He went to France and Sweden to examine hydroelectricity plants and hydrogeology in 1958 and explored his interest in cave sciences even further, contacting speleology organizations in these countries. He published his first book on caves in Turkey and caves of the world, the following year (Aygen, 1959). This is the first book in the field of speleology in Turkey, and it describes caves like Maraspoli Cave, Karagöl Sinkhole, and Damlataş Cave which were explored and surveyed by Aygen.

Fig. 1.6 Temuçin Aygen in Narlıkuyu Cave (Aygen, 1984)

Other than this work, Aygen had six published books, 92 articles, and 11 presentations in various international conferences (Erdem, 2013). Following the establishment of MAD in 1964, Temuçin Aygen invited French cavers to Turkey and together they explored hundreds of caves throughout many different regions of Anatolia since 1965.

1.2 Collaboration with European Associations

In the period following the establishment of MAD in 1964, Temuçin Aygen contacts foreign cavers and scholars for Turkey's speleological opportunities to be researched as scientifically as possible. It is highly possible that both studying in Switzerland and the insufficiencies he detected during his 10 years of studying caves in Turkey led him to decide to work with foreigners.

The first serious contact with foreign science people was made at the International Conference of Speleology and Karstology which was organized in Istanbul between the 24th and 26th of September 1964. Over 20 globally well-known foreign scholars on the field of speleology like Norbert Casteret, Paul Fenelon, Joseph Newell Jennings, Claude Drogue, Claude Pommier, and Kuchta Gijula attended this conference. After the three-day conference, attendees took a field trip across Anatolia through Ankara, Konya, Mersin, Antalya, Burdur, Izmir, and Bursa for 2 weeks. Thus, they had the opportunity to investigate the principal karstic phenomena of Turkey: Konya dolines, travertines of Yerköprü and Antalya, caves of Mersin region, vauclusian springs of Manavgat. The interest of these occurrences of Anatolian karst is unquestionable; in addition to the scientific problems they pose, they represent a great economic value either as tourist centers or as producers of electric energy and sources of water for irrigation (Fenelon, 1968).

On the other hand, again with the invitation of Temuçin Aygen, foreign cavers came to Turkey and began their cave studies in 1965. Three different SCP member teams researched the vicinity around Konya–Antalya–Beyşehir. In this first research, previously known caves such as Maraspoli, Yerköprü, Insuyu, and Cennet–Cehennem were examined and surveyed again. Some of the answers to Aygen's cave survey forms, which were previously sent to villages, were analyzed in this same study. This led to the discovery of Pınargözü Cave and mapping of two large caves: Yalan Dünya and Toy Islam (Couderc, Pelon, & Conrau, 1965). The team which came the following year started studying the caves and aquifers around Manavgat Gorge, which completely consists of limestone, for the construction of the Oymapınar Dam. Dumanlı Spring, which resides on the eastern shore of this gorge, was submerged under the dam lake waters after the construction and has an outflow of 50 m^3/s in the high season. It was tried to be reached from the caves above Dumanlı and some other aquifers. The challenging Altınbeşik Cave, which resides west of the Manavgat River, was discovered and surveyed the same year (C. Chabert, 1966). Altınbeşik Cave was re-surveyed and remapped three times more in the following years. French cavers' studies with Temuçin Aygen around Antalya–Manavgat River continued in 1967. Düden, Kelebekli, Güvercin, Koyungöbedi, Çayırönü, and several other caves were explored and surveyed. Michel Bakalowicz's hydrogeological research around Manavgat at that time would later be his Ph.D. thesis (Bakalowicz, 1967; C. Chabert, 1967). 1968 and 1969 were the years SCP worked in both Manavgat Region and Pınargözü Cave. Pınargözü Cave's research was still under progress in 1968. The cave had reached 1800 m in length and +30 m of height. Many other caves like Zindan, Tınaztepe, Değirmenlik, and Düdencik were also surveyed. Among these, with a depth of −330 m, Düdencik Sinkhole was the deepest cave in Turkey for 30 years until the beginning of the Çukurpınar Sinkhole research in 1989 (Bakalowicz, 1968).

French cavers discovered, surveyed, mapped, and published 74 caves in the first five years they worked in Turkey. More importantly, they found big and important caves like Pınargözü, Tınaztepe, and Altınbeşik and they conducted geological and hydrogeological studies in detail in the Mid-Taurus Range.

Spanish cavers went through Ayvaini Cave from end to end and drew a rough map of this phreatic traverse cave in 1970 (Fig. 1.7) (Agnoletti, Baldieri, Fiorentini, & Ortensi, 1970). A highly detailed map of this cave would be drawn by Claude Chabert many years later. SCP, CMN and English cavers reached 3220 m in length and +190 m of height in Pınargözü Cave in a joint venture in the same year. The heat in the cave was 10.5 °C, and water had a temperature of 6 °C in August 1970. It had countless waterfalls to be climbed and 3 siphons which were required to be dived in (Fig. 1.8). It had become so hard for the French team to work in the Pınargözü Cave, and they started searching for a sinkhole on Dedegöl Mountain to reach the cave from above. In this first year of the work, which would take many more years, 8 different sinkholes were found and examined but it was understood that none of them had a connection to the Pınargözü Cave.

Explorations around the Manavgat Gorge were under progress the same year; Yedi Miyarlar aquifers were being studied, and Susuz Cave, which was discovered the previous year, was explored (Bakalowicz, 1970; C. Chabert, 1970) . Two different highly crowded teams which consisted of French and English cavers reached 5275 m in length and +248 m of height in Pınargözü Cave at the end of their studies in 1971 and 1972. The cave was not completely studied, but the survey which began in 1965 was finalized after seven years, in 1972. The ongoing search for a connection on Dedegöl Mountain was left without a result, even though five other sinkholes were found. Other than Pınargözü Cave, 23 caves were studied and 14 caves were mapped during these years. Important caves like Susuz Cave, Eskiyörük Cave, Ferzine Cave, and Gölcük Sinkhole were among these (C. Chabert, 1972a, 1972b) . A total of 131 caves were discovered and explored by French cavers between 1965 and 1972.

Previously found 25 caves, some of which were not completely examined, were surveyed in 1973. Among these, the survey and mapping of Kayaağıl Doline, which is the biggest doline in Turkey, and Felengi Cave, are important. Kayaağıl Doline is a huge collapsed doline, which has 500 m on its long edge and −160 m of depth, with a forest at its ground (Dobrilla, 1977). SCP and CMN continued

Fig. 1.7 First traverse of Ayvaini Cave (Agnoletti et al., 1970)

Fig. 1.8 In front of the first waterfall at Pınargözü Cave during the fourth year of exploration, 1970 (Photograph C. Chabert)

studies in the Dedegöl Mountain and Manavgat River Basin in 1974 (C. Chabert, 1975) and further investigations on the west block of Manavgat River in 1976. Kembos and Eynif poljes, and sinkholes around the vicinity were measured and mapped. Fifteen different caves were studied in 1976, and the most important activity of this season was the discovery and measurement of Tilkiler Cave (Fig. 1.9). It was measured to be 2755 m on this first survey of the cave (C. Chabert, 1976). On Temuçin Aygen's call, SCP and CMN go to Black Sea Region in 1977 to research caves that Aygen discovered with English cavers between 1975 and 1976. There were nine caves around Zonguldak which were previously known but not surveyed. Among them were long caves previously known and entered by Spanish cavers (Masriera & Martorell, 1971) and Aygen: like Gökgöl, Cumayanı, Kızılelma, and Ilıksu caves. Along with some other small caves, Cumayanı Cave's first survey and mapping were conducted on this study (C. Chabert, 1977). After this research in the Black Sea Region, they returned to the Taurus Mountains. Over 20 sinkholes and caves were discovered and surveyed in Avason and Sevinc districts. New caves were found in Manavgat–Oymapınar Dam area, and studies in Tilkiler Cave were continued. After measuring, this cave was revealed to be 4845 m and became Turkey's longest mapped cave. Another interesting activity of the same year was French cavers' re-survey of the Kocain Cave even though it is not in the same region. This enormous chamber researched by Temuçin Aygen in 1972 is measured and mapped again 55 years after Moretti. It can be said that the year of 1977 is the most successful one of the 12 years for SCP members and other French cavers since they began research in Turkey in 1965, not in the numeric sense but the aspect of productivity (C. Chabert, Callot, Chabert, & Gilli, 1978). The following year 5585 m was reached in the Tilkiler Cave, and it became one of the longest conglomerate caves of the world. Furthermore, some small sinkholes like Akpınar, Tefekli, Erkibet, and newly found caves around the Manavgat River were surveyed and mapped. The total of caves researched in 1978 was 15 (J. Chabert, 1979; Gilli, 1979). Trent Polytechnic Speleological Society working in the Black Sea Region surveyed and mapped the Kızılelma Cave found in Zonguldak; total length of the cave was 6250 m, and a possible connection with Cumayanı Cave had been found. 2970-m-long Gokgol Cave, 2725-m-long Bulak Mencilis Cave, and 1085-m-long Cumayanı Cave were also

Fig. 1.9 Big Lake of Tilkiler Cave, 1976 (Photograph E. Gilli)

Fig. 1.10 Sakaltutan 1 Cave, 1979 (Photograph E. Gilli)

surveyed in this work of Trent Polytechnic (Watkins, 1980). These four caves were re-surveyed and remapped, and new branches of them had been found in the following years. French cavers surveyed over 1000 m more in Tilkiler Cave in 1979, and it became the second-longest conglomerate cave of the world with a total length of 6600 m. The same year, two sinkholes were discovered in Süleymaniye Village near Akseki. The research on these two sinkholes named Sakaltutan no. 1 and no. 2, which are very close to each other, would be completed the following year (Fig. 1.10). As an interesting coincidence, Sakaltutan Sinkhole no. 1 is -303 m and Sakaltutan Sinkhole no. 2 is −302 m. Following the Düdencik Sinkhole, these two sinkholes then became the second and third deepest caves in Turkey (J. Chabert, 1980). The article "Les Plus Grandes Cavités Turques" written by Claude Chabert, which was published in 1981 in Grottes et Gouffres, includes a list of longest and deepest caves explored in Turkey until 1980 (C. Chabert, 1981). Of the 13 longest caves on the list, six were explored by French, five were explored by English, and two were explored by Spanish and BÜMAK cavers, respectively. The longest cave on the list was Tilkiler Cave with a total length of 6600 m. With its newfound branch, Tilkiler Cave's length reached 6818 m, and it is the third-longest cave on the list of longest caves of Turkey today. There are 25 caves on Chabert's deepest cave list of 1980; four of them were explored by English, two of them were explored by BÜMAK, one of them was explored by Spanish, and the rest of them were explored by French cavers. The deepest cave on this list was the Düdencik Sinkhole with the depth of −330 m, which now became the 21st deepest cave after 40 years in between.

1.3 Turkish Caving

MAD is the first speleological society of Turkey, which was established by Temuçin Aygen in Ankara in 1964. The society, which in its first 15 years following its establishment could not get the chance to make speleological research independently and increase the number of its members mainly because of Aygen's collaboration with foreign and especially SCP member cavers, improved drastically in the 1980s. The establishment, which got stronger with the addition of its new members after 1985, announces its studies through the MAD Bulletin, which published 16 issues until today since 1987. Even if we exclude the studies it conducted with foreign organizations between 1964 and 1987, MAD has surveyed and mapped 167 caves since 1987. These include deep and important caves like Subatağı Sinkhole (−643 m), Sütlük Sinkhole (−640 m), Düdenağzı Sinkhole (−612 m), and Kocadağ Sinkhole (−458 m). Their explorations on Yaylacık Cave System, which is found in Mid-Taurus Mountains and connected to two different sinkholes, continue. With its 5929 m length and −595 m depth, this cave is one of the biggest cave systems in Turkey. The establishment mainly works in Bursa and Kütahya provinces and Mid-Taurus Range.

During the 2nd National Speleology Symposium organized in 1994, MAD ushered in the establishment of the Speleological Federation of Turkey (TMF) and led the way to the establishment of a rescue team within the constitution of this federation. Beginning with the year 2000, MAD organizes rescue training and the annual joint training that all the speleological associations active in Turkey could attend. With the four different cave documentaries it shot to be shown on Turkish National TV channels, MAD also supported the establishment of new university speleological clubs in recent years.

BÜMAK, which was established in 1973, nine years after MAD, is the first university cave club in Turkey. Activities of the club, which were limited to caves around Istanbul in the years following its establishment, began to spread to Turkey with the explorations in the Taurus Mountains in 1976 and Bulak Mencilis Cave in 1977. Its activities accelerated with the very first crossing of the Ayvaini Cave in 1978, discovery and publishing of the 3150-m-long Dupnisa Cave (Atalay & Ülkümen, 1980) and the discovery of 10 new caves during the preliminary survey in Antalya–Akseki–Çimiyayla. During the Çimiyayla Expedition conducted with the Imperial College Caving Club in 1979, BÜMAK cavers used Single Rope Technique (SRT) for the first time and completed the survey of 25 caves (Fig. 1.11) (Yamaç, Aktar, & Atalay, 1980).

Beginning in 1980, BÜMAK started block research in Kastamonu Province. Over 20 deep and long caves were explored in this area within three years. Explorations in Muğla, Izmir, and Bursa provinces were also being conducted around the same time. Report of a deep cave within the Taurus Mountains in 1989 led to BÜMAK's exploration of −1192-m-deep Çukurpınar Sinkhole and −1429-m-deep Peynirlikönü Sinkhole. After the conclusion of the explorations on these two sinkholes, BÜMAK directed its surveys on the uninvestigated parts of the Taşeli Plateau. Over 20 caves were found, surveyed, and mapped on the research conducted north of Gazipaşa (Yamaç, 2003).

Fig. 1.11 Dünekdibi Pit (−197 m), 1979 (Photograph A. Yamaç)

Organizing the 1st National Speleology Symposium in 1990 in Istanbul, BÜMAK created an opportunity for Turkish cavers to truly meet each other for the first time and laid the groundwork for TMF, which would be established four years after this date. Society continues its work in many different regions of Turkey. Examining BÜMAK's journal called Delta, which published eight issues until today, it can be seen that they had surveyed and mapped 101 caves in the 16 years between 1978 and 1994 and this number increased to 152 in the second 16 years between 1995 and 2011. On the other hand, they worked in four regions in the first 16 years and this number increased to eight later on. BÜMAK has eight caves within the deepest 25 caves and six caves within the longest 25 caves of Turkey. The biggest structural problem BÜMAK has is the graduation of its members from the university within four years and their departure from the club. To present a solution to this problem, old members who wanted to continue as cavers established BUMAD in 2007.

Cave Exploration Unit established within the constitution of the General Directorate of Mineral Research and Exploration (MTA) in 1978 is the third cave organization in

Turkey. This establishment published 62 reports and researched over 900 caves in the past 40 years. All the teams in their studies consist of experts, which makes these reports scientifically detailed. Compared to other establishments focusing on speleology in Turkey, which of many are sportive, MTA's studies are considered to be the most important scholarly studies conducted in Turkey. Survey and mapping of many big caves like Kızılelma, Cumayanı, Gökgöl, and Sofular and discovery of the Kuzgun Sinkhole are among the most important researches of this organization. One of the two important problems this unit encounters, who conducted many block surveys in areas like Bolkar, Aladağ, Siirt, Zonguldak, Manavgat, and Thrace regions, is not being independent and working according to requests and directives because of its status as a governmental organization. In a country like Turkey where studying and surveying still requires serious procedures, MTA, which has no obstacle both permission-wise and financially, uses some of its resources and labor to arrange some caves as show caves according to demands and cannot be effective in terms of protection. Another structural problem they encounter is that they cannot collaborate with any caving clubs or societies because every member of this division is a government official. On the other hand, a joint venture as such would provide for other establishments which have a lack of scientific speleological knowledge.

Following these first three establishments, 15 different speleological clubs and associations were established in Turkey since 1978. Even though this quick rise in numbers, collaborations in recent years caused many establishments who are active in the same area to work together. One of the main reasons for these collaborations was the joint activities organized by TMF where cavers of various clubs got to know each other and started to collaborate, which lessened the competition. Furthermore, as in the example of BÜMAK and BUMAD, other universities' cave club members also established new societies after their graduations or joined other societies, causing cavers from different backgrounds to become closer. It can be said that competition in terms of exploration continues at some level but it is productive for Turkish speleology, not harmful.

The establishment of TMF which had begun being debated over in 1992 and announced with a joint declaration of MAD and BÜMAK during the 2nd National Speleology Symposium in 1994 in Ankara was accepted by other groups that attended and started its activities the same year. The prepared and approved first regulation proposal included articles on the federation's definition, purposes, membership procedures, executive council, working principles, and missions. Following its establishment, TMF became a member of the International Union of Speleology (UIS), European Speleological Federation (FSE), and Balkan Speleological Union (BSU). Other than the rescue training that all establishments attend, the federation also organizes joint caving activities. The Balkan Cavers meeting, which is organized by BSU in a different Balkan country each year, was organized by TMF in Turkey twice, gathering hundreds of cavers from ten different countries and enabling them to exchange information and perform caving together.

Even though Turkish caving has improved in the past 30 years, various problems appeared with these improvements. "Deep cave enthusiasm" is one of those problems. Following the discovery of Çukurpınar and Peynirlikönü sinkholes, which had an important place on the list of deepest caves of the world in the years they were discovered, Turkish cavers who especially focus on sportive caving and young cavers who are members of university clubs started searching for more deep caves. Besides the facts that long caves require horizontal bedding and this type of geological formations are rare in Turkey, the requirement of time for wide-area explorations, an increased delicacy for long surveys, and more experienced teams to dive into encountered siphons also make deeper caves more preferable to longer caves in these types of explorations.

On the other hand, "cave depth record sickness" that increased over the 1990s ushered by Russian and Ukrainian cavers and the fact that Turkey has more deep caves instead of long caves led many foreign and Turkish societies to this type of cave explorations. These establishments lack the resources for large field researches, which requires time and financial support in Turkey, where karstic areas are extremely wide. The requirement of preliminary studies is another problem encountered while planning for this type of wide-scale field research. There are not many karst experts in Turkey who can give scholarly support to cavers that intend to conduct such field research. Because of the lack of an institution or an organization that conducts scientific speleology research, some science people hold a monopoly on this subject. Limited to a few books and articles, scientific cave and karst studies are far from creating a general context and helping cavers who want to conduct research.

Under this speleological attitude, instead of focusing on a particular area and researching that area over a long period, establishments work disorderly on different areas according to the arriving news and reports, which is another mistake being made continuously for many years. This continuing error combined with insufficient database and reference usage causes different caver groups to research the same areas, and thus many caves are being re-explored and remapped. Fifty-eight caves are detected to have double maps until now, and some of these are large caves (Yamaç, 2013). "Double maps" in this sentence is used as a term, as some caves have three, four even five different maps. Even if the database would be used, this problem would continue as long as field research is made without an extensive bibliography search and previously explored −233-m-deep sinkhole named G2 in the Taurus Mountains will undoubtedly be re-explored and re-surveyed a few more times.

Interestingly, the final problem is encountered after overcoming all these problems, and going to the field is socioeconomical. Because of rural to urban migration and the decline of sheep farming in the past 20 years, it caused the almost complete disappearance of shepherds in Turkey. Cavers used to be informed about deep caves by local villagers and shepherds during explorations in the past. Reconnaissance trips in mountainous areas became much more time consuming and difficult without these opportunities.

Considering all these problems, it would be more beneficial for Turkish caving to head toward a method that sees different teams working together, focusing on a specific area, examining it step by step with the support of geological and geomorphological scholars, instead of just evaluating reports from local people. Joint ventures observed in recent years are good examples of this type of survey.

References

Agnoletti, P., Baldieri, A., Fiorentini, A., & Ortensi, P. (1970). Campagna Speleologica in Turchia. *Notiziario del Circolo Speleologico Romano, XV*(20–21), 5–37.

Altay, B. (2007). Temuçin Aygen'in Ardından. *MAD Bülteni, 13,* 68–72.

Atalay, E., & Ülkümen, O. (1980). *Kırklareli-Dupnisa Mağara Sistemi Araştırma Raporu.* Retrieved from İstanbul.

Aygen, T. (1959). *Mağaralar ve Yeraltı Irmakları.* Ankara: DSI.

Aygen, T. (1984). *Türkiye Mağaraları.* İstanbul.

Aygen, T. (1988). *Türkiye'nin Az Bilinen Doğal ve Arkeolojik Değerleri.* İstanbul.

Bakalowicz, M. (1967). Hydrogéologie du Taurus Calcaire Entre Beyşehir et Manavgat. *Grottes et Gouffres, Bulletin Périodique du Spéléo-Club de Paris, 40,* 29–37.

Bakalowicz, M. (1968). Données Géologiques, Hydrologiques et Météorologiques sur les Cavités, Pertes et Émergences Reconnues Pendant la Campagne 1968. *Grottes et Gouffres, Bulletin Périodique du Spéléo-Club de Paris, 42,* 27–42.

Bakalowicz, M. (1970). Campagne Spéléologique 1970 en Taurus Occidental Expédition française en Turquie. *Grottes et Gouffres, Bulletin Périodique du Spéléo-Club de Paris, 45,* 15–27.

Bey, A. (1869). Die Umgebung des See's Kütchücktschekmetché in Rumelien. *Verhandlungen der k.k. Geologischen Reichsanstalt, 12,* 263–265.

Bey, A. (1872). Études Géologiques sur les Environs de Constantinople, Yarym Bourgas, Macri-Keuy, Sari Keuy. *Gazette Médicale d'Orient, Constantinople, Octobre, 7,* 103–105.

Bousquet, R. (1901). Les Grottes de Yarim-Bourgas. *Échos d'Orient-Stamboul, 4,* 295–302.

Chabert, C. (1966). Turquie 66. *Grottes et Gouffres, Bulletin Périodique du Spéléo-Club de Paris, 38,* 2–21.

Chabert, C. (1967). Lettre de Turquie. *Grottes et Gouffres, Bulletin Périodique du Spéléo-Club de Paris, 40,* 5–14.

Chabert, C. (1970). Pınargözü ou le Spéléologue Provoqué. L'expédition Spéléologique Franco-anglaise dans le Taurus 1970. *Grottes et Gouffres, Bulletin Périodique du Spéléo-Club de Paris, 45,* 3–14.

Chabert, C. (1972a). Cavités de la Campagne 1971. *Grottes et Gouffres, Bulletin Périodique du Spéléo-Club de Paris, 48,* 43–46.

Chabert, C. (1972b). De l'expédition à Pınargözü en l'an 1971. *Grottes et Gouffres, Bulletin Périodique du Spéléo-Club de Paris, 48,* 5–28.

Chabert, C. (1975). Enigmes turques. Du massif du Dedegöl dağı aux gorges de la Manavgat (Taurus). *Grottes et Gouffres, Bulletin Périodique du Spéléo-Club de Paris, 55,* 3–11.

Chabert, C. (1976). La campagne 1976 dans le Taurus. La Cavité de Grande Promesse. *Grottes et Gouffres, Bulletin Périodique du Spéléo-Club de Paris, 62,* 3–24.

Chabert, C. (1977). Les Grottes de la Mer Noire (Zonguldak, Turquie). *Grottes et Gouffres, Bulletin Périodique du Spéléo-Club de Paris, 66,* 3–13.

Chabert, C. (1981). Les Plus Grandes Cavités Turques. *Grottes et Gouffres, Bulletin Périodique du Spéléo-Club de Paris, 79,* 17–23.

Chabert, C., Callot, Y., Chabert, J., & Gilli, E. (1978). Les recherches de 1977 sur le système hydrologique d'Eynif (Taurus occidental, Turquie). *Grottes et Gouffres, Bulletin Périodique du Spéléo-Club de Paris, 67,* 3–28.

Chabert, J. (1979). La campagne 1978 en Turquie (Manavgat, Antalya). *Grottes et Gouffres, Bulletin Périodique du Spéléo-Club de Paris, 71,* 3–22.

Chabert, J. (1980). L'expédition 1979 dans les Monts du Taurus (Turquie). *Grottes et Gouffres, Bulletin Périodique du Spéléo-Club de Paris, 75,* 3–12.

Couderc, M., Pelon, J., & Conrau, (1965). L'expédition 1965 en Turquie. *Grottes et Gouffres, Bulletin Périodique du Spéléo-Club de Paris, 36,* 3–47.

Dobrilla, J. (1977). Expédition Turquie 1973. *Grottes et Gouffres, Bulletin Périodique du Spéléo-Club de Paris, 65,* 16–24.

Erdem, B. (2013). *Türkiye Speleoloji Kaynakçası.* Istanbul.

Fenelon, P. (1968). Compte-rendu de la Conférence Internationale de Spéléologie et de Karstologie d'İstanbul (24 sept.–14 oct.1964). *International Journal of Speleology, 3*(1–2), 71-88.

Gilli, E. (1979). Expedition Eynif 1979. *Spéléologie. Bull. Club Martel Nice, Club Alpin Français, 105,* 10–27.

Hovasse, R. (1927). Yarımburgaz Mağarası (La Grotte de Yarım Bourgas). *Istanbul Darulfününü Fen Fakültesi Mecmuası, 5,* 395–422.

Kadioğlu, S. (2003). Raymond Hovasse'ın Türkiye'deki Bilimsel Çalışmaları Ve Baltalimanı Hayvanat İstasyonu'nun Kuruluşu. *Osmanlı Bilimi Araştırmaları, 4,* 61–81.

Kansu, Ş. (1939). Türk Tarih Kurumu Prehistorik Araştırmaları. *Belleten, 3*(9), 93–97.

Kocacan, H. (1918). Bir Mağara Nasıl Tetkik Olunur ve Yarımburgaz Mağarasını Ziyaret. *Tedrisat Mecmuası, 61,* 12–18.

Martel, E. (1911). *Notice sur les travaux scientifiques.* Paris Masson edit.

Masriera, A., & Martorell, J. (1971). Memoria de la expedición espeleológica Turquía 1970 del Club Montañes Barcelonés. *Karst, 8*(29), 15–25.

Moretti, G. (1923–1924). In Daghindà Quogia In, La Grande Caverna nella Montagna delle Caverne. *Annuario della Reale Scuola Archeologica di Atene, 6–7,* 509–546.

Viale, V. (1925–26). Relazione sull'attiva della Missione archeologica di Adalia nell'anno 1922. *Annuario della Reale Scuola Archeologica di Atene, 8–9,* 357–392.

Von der Osten, H. (1929). *Explorations in Hittite Asia Minor 1927–1928.* Chicago: University of Chicago Press.

Watkins, P. (1980). *Trent polytechnic speleological expedition to Turkey 1978.* Retrieved from Nottingham.

Yamaç, A. (2003). *Boğaziçi Üniversitesi Mağara Araştırma Kulübü'nün 30 Yılı (1973–2003).* Istanbul.

Yamaç, A. (2013). *Double Cave maps of Turkey.* Paper presented at the 16th International Congress of Speleology, Brno.

Yamaç, A., Aktar, M., & Atalay, E. (1980). *Çimiyayla Araştırma Raporu.* Retrieved from Istanbul.

2 Cave Archaeology in Turkey

Abstract

We have comprehensive reports on archaeological surveys and excavations conducted in a total of about 90 caves in Turkey. There were several more studies conducted in the 1940s, but the publications of these researches are uninformative. Even though there are archaeological excavations conducted in a limited number of caves during the recent years, the results are striking. Numerous tools dating back to tens of thousands, even hundreds of thousands of years were found and important archaeological explorations were made in not just the important excavations conducted in long periods that were explained in Chapter 5, but also in excavations of smaller scales. Interestingly, no hominid or Neanderthal skeletal remains were found in almost none of the aforementioned excavations, except a few small pieces. These smaller-scale cave excavations that have given various kinds of findings are explained in this chapter.

2.1 Introduction

In almost every map that shows early human migrations, there is often one arrow showing a hypothetical route through Anatolia and Thrace toward Europe. Due to its location at the junction between the European and Asian continents, Anatolia is one of the theoretical migration routes of the hominins that came out of Africa (Dinçer, 2016; Otte, 1998). Even though Anatolia is expected to be a bridge between the early human findings of Europe, Asia, and Africa due to its location between these continents, this relation could not be properly searched until now mainly due to low popularity of Paleolithic archaeology in general and cave archaeology in particular (Harmankaya, 1997). Today, researches are executed by a handful of devoted science people and the limited data we have left many questions unanswered on Paleolithic culture in Anatolia and early human migrations between the continents. With those limited findings, it is impossible to make a connection between the cave excavations conducted in European, Asian, and Middle East countries and to come up with a conclusion about Anatolia's role in early human migrations. Future cave excavations in Anatolia may be expected to provide both solid facts about the Paleolithic Period of the area and new data on early humans' migrations toward Europe and Asia.

On the other hand, even the limited number of archaeological cave surveys and excavations conducted after the 1990s in Turkey, which is known to have thousands of caves, shows us that the findings in these caves are not just limited to Paleolithic Period but also reaches to Neolithic Age, Bronze Age, and even Classical Periods. But, if we exclude a few scholarly studied caves explained in detail in Chap. 4 which have classical importance like Zindan and Kocain, most cave surveys and excavations in Turkey deal with findings from Paleolithic and Neolithic ages.

However, apart from the artificial caves in Cappadocia, historical remnants that date back to the Classical Period like inscriptions, chapels, and churches can be found in many natural caves and academic surveys on this type of caves are insufficient. There is not any comprehensive research on cave churches like Içel–Cennet, Antalya–Gedifi Ini, and Antakya–St. Pierre. There are neither studies nor articles about the chapel with an apse and several other structures carved on the rocks inside the Yarımburgaz Cave.

First archaeological cave surveys of Turkey were carried out by Italian archaeologists Giuseppe Moretti and Vittorio Viale around Antalya. In his two different studies, Moretti copied the inscriptions inside Kocain and Karain caves (Fig. 2.1) (Moretti, 1923–1924a, 1923–1924b). Even if he mentioned it as the "Sacred Cave of luvadja," we know that this cave is the Karain Cave near to Yağca Village. Moretti gave detailed information about this cave in his article. In the study he conducted in a cave found in Gurma Village 20 km west of Antalya, Viale opened a 2 × 3 m test trench and

A. Yamaç et al., *Caves and Karst of Turkey—Vol. 1*, Cave and Karst Systems of the World,
https://doi.org/10.1007/978-3-030-65501-3_2

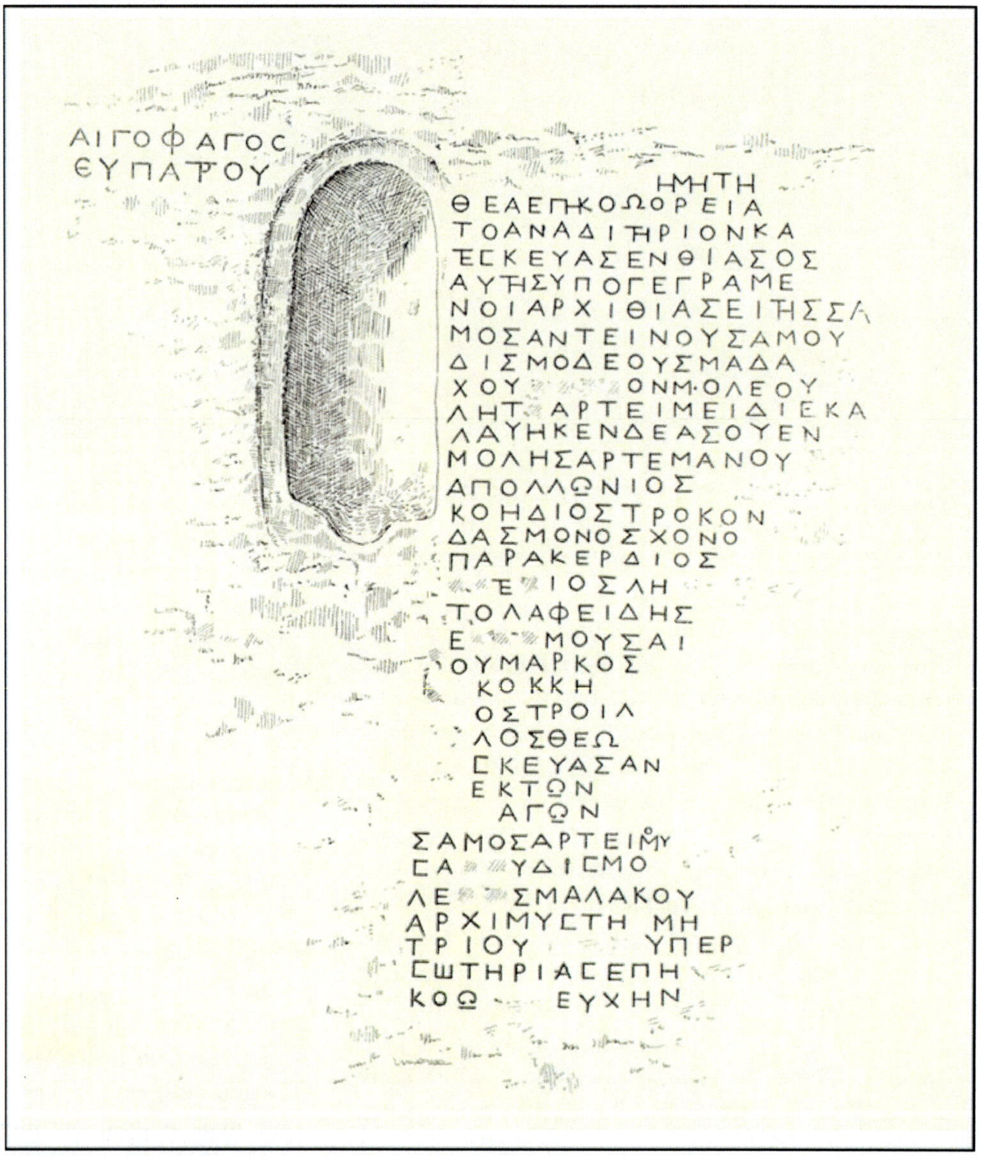

Fig. 2.1 Roman inscription from the outer wall of Karain Cave (Moretti, 1923–1924b)

found Paleolithic and Neolithic microliths and pieces of pottery in different strata (Viale, 1925–26).

Following these first surveys and excavations, from the 1940s and onward, archaeological cave surveys started to be organized more often. Following the academic tradition of that period, almost all of these studies were conducted by anthropologists, mostly by Şevket Aziz Kansu, Enver Bostancı, and Kılıç Kökten. Especially Kökten's surveys and excavations conducted until the 1970s all around Turkey led to the discovery of many cave settlements. On the other hand, examining their articles and reports, it can be said that the excavations in this period rather test trenches compared to cave excavations in the modern sense and their publications were insufficient and unscientific. From the 1920s and onward, there have been archaeological surveys and/or excavations in numerous caves in Turkey (for a comprehensive list and bibliography till 1994 ASPRO (Hours et al., 1994)). Even if these early-period cave excavations are insufficient in terms of scientific content, the cave excavations conducted in collaboration with international cooperations after 1980, which included applications from different branches of archaeology, provided remarkable results. However, the scientificity of these excavations and the limited number of academic personnel resulted in a very small number of cave excavations in Turkey in the past 40 years and during the last few years, there are 5 or 6 cave excavations. Some archaeologists who are interested in cave archaeology followed the works realized before them and surveyed mostly in Mediterranean and Marmara regions, while there are almost no surveys in other regions that are known to have thousands of caves.

On the other hand, serious developments in the field of archaeological cave inventory are being monitored in recent years. The first known inventory study on the archaeological evaluation of caves in Turkey was executed in 1951. Ş.A. Kansu's survey at the date, which was sent to all provinces to detect caves and rock shelters in Turkey, was emphasized to be unhealthy by Kökten (1952; Taşkıran, 2018). According to this rough inventory results dated back to 1951, there are 3534 artificial and 5952 natural caves in Turkey. From 1990s and onward, comprehensive cave inventory surveys started to be conducted in different regions in Turkey with participation of experts (e.g., Güleç, Özer, Sağır, Baykara, & Şahin, 2013; Özdoğan & Karul, 2002; Taşkıran & Kartal, 2004; Yalçınkaya, Kösem, Özçelik, Erek, & Kartal, 2000) and all these surveys are being added to The Archaeological Settlements of Turkey Project database (www.tayproject.org) through annual updates (Harmankaya & Tanındı, 1997; Harmankaya, Tanındı, & Özbaşaran, 1997).

Cave paleontology surveys and excavations are completely left under the shadow of archaeological excavations. Although some paleontological excavations are being conducted in 7–8 different locations of Turkey today, in cave excavations it does not go further than the examination of animal skeleton fragments found. Still, there are some notable examples like Ursus *deningeri* and Ursus *spelaeus* bones found in Yarımburgaz Cave excavations (Fig. 2.2) (Arsebük, 1998).

It is important to make a short explanation about the "Paleolithic Period" term and its chronology which is frequently used in this and the following chapters. This archaeological period is examined in three subdivisions: Lower, Middle, and Upper Paleolithic. The starting date and subdivision dates of this period are the subject of an ongoing discussion and differentiate each continent. Especially Levant's Paleolithic chronology is notably different from that of Europe's. We prefer taking J. J. Shea's book "Stone Tools in the Paleolithic and Neolithic Near East" as a reference for the table in Fig. 2.3 (Shea, 2013).

Fig. 2.2 Skull of Ursus *deningeri* from Yarımburgaz Cave (Photograph G. Arsebük)

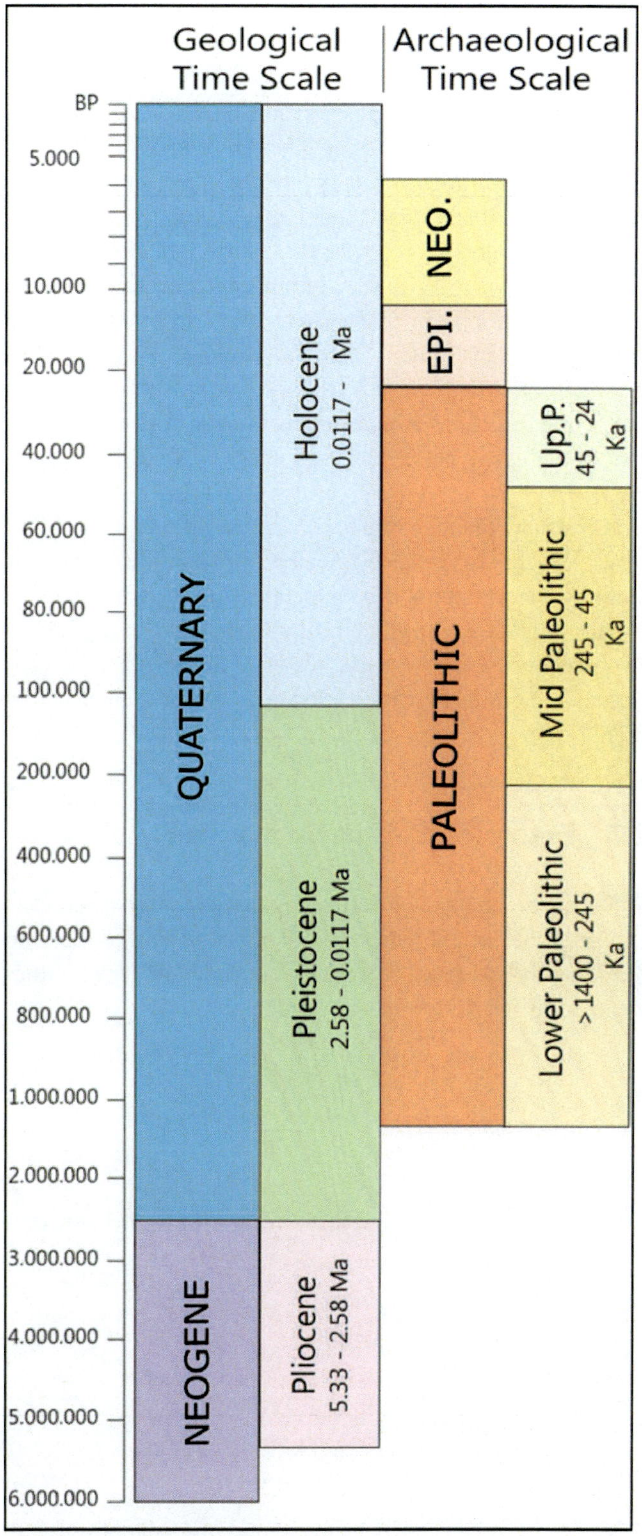

Fig. 2.3 Geological and archaeological timescales, dates, and periods are adapted from Shea (2013)

"Paleolithic Period," the longest period in archaeology, begins with the first stone tool production of the hominins and ends with 24,000 BP. The transition period between the end of the Paleolithic Period in 24,000 BP and the start of the Neolithic Period in 12,200 BP is named Epipaleolithic Period. This period, which corresponds to the Mesolithic Period in European archaeology, is accepted as a transition period between hunter-gatherer humans and Neolithic communities.

All the dates in this chapter and onward are given as Before Present (BP), and these dates are based on C14 tests made on archaeological layers. On the other hand, in this type of cave excavations, there may not be any organic material to make a C14 test on; thus, stone tools found in the excavations may be dated through their comparison to similar tools found in other excavations. In situations like this, findings are roughly dated and referred through their specific archaeological period. Any date mentioned in this and following chapters and any date given for these subdivisions which are not certain are subject to the dates given in Fig. 2.3.

Karain, Öküzini, Suluin, Yarımburgaz, and Üçağızlı caves which are archaeologically important due to their findings and results will be explained in detail in Chap. 4. Archaeological findings of Kadıini and Kızılin caves are also mentioned in the same chapter. Some other archaeologically significant cave excavations are explained below (Fig. 2.4).

2.2 Examples of Cave Excavations

Some of the archaeological cave excavations or researches explained below are rather old but have striking results. On the other hand, few others are comparatively new, rather small scale but with important findings.

2.2.1 Beldibi Rock Shelter

This small rock shelter located south of Antalya, 3 km north of the village of Beldibi, is thought to be used as a temporary campsite or as a habitation area by Upper Paleolithic, Epipaleolithic, and Neolithic hunters. There are paintings on the back wall of this rock shelter. Excavations here, with an initial trench by the wall paintings, began in 1959, three years after the site was discovered by Bostancı (1959). Excavations were continued in 1960, 1966, and 1967. The depth of the first trench opened was 4.50 m at the deepest point. Epipaleolithic and Upper Paleolithic microliths and chipped stone tools were found at the bottom layer in these excavations. A pebble painted entirely red with iron oxide paint, a pierced amulet of sandstone, several seashells, deer antlers and teeth, and human skull fragments were found in the Upper Paleolithic Stratum. Another pebble found in Upper Paleolithic Stratum is also reported to have been shaped like a human face in profile (Bostancı, 1966). It is not clear from the publications whether virgin soil has been reached (for a schematic stratification of the site, see the table in Bostancı (1967b)). Perhaps the rock shelter's most important element is the rock art. While most of the paintings have been made of red iron oxide paint, some have been engraved. The engravings are of deer figures outlined along the natural veins of the rock face with a few simple lines. This antlered animal is portrayed as fleeting with its head turned back. Bostancı dates this work of art to the Upper Paleolithic (Bostancı, 1959).

2.2.2 Belbaşı Rock Shelter

The site is located 7 km northwest of Beldibi Village, near Beldibi Rock Shelter. It was discovered in 1959, and E. Bostancı started excavations the following year. A total of 13 different layers were excavated in a total depth of 160 cm

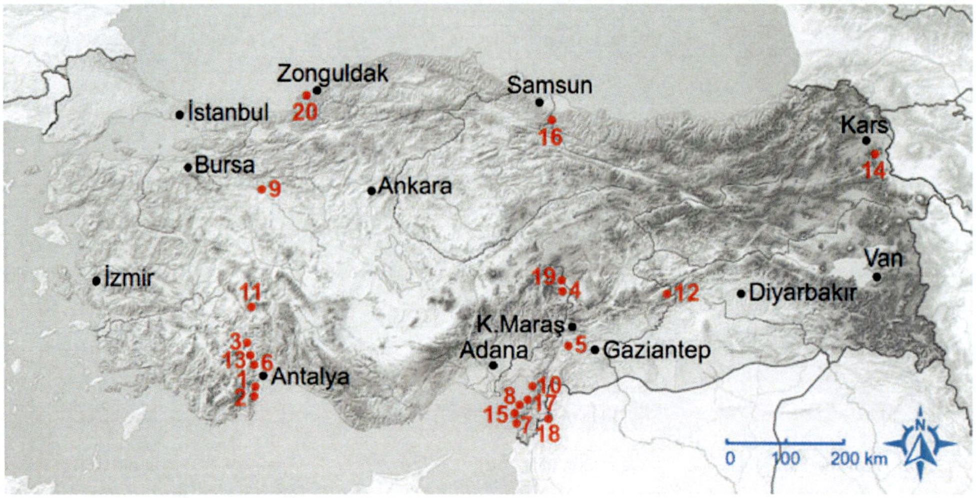

Fig. 2.4 Locations of 20 cave excavations mentioned in this chapter: 1 Beldibi, 2 Belbaşı, 3 Çarkini, 4 Direkli, 5 Dülük, 6 Gurma, 7 İkiağızlı, 8 İncili, 9 İnönü, 10 Kanal, 11 Kapalıin, 12 Keçiler, 13 Kızılin, 14 Kurbanağa, 15 Merdivenli, 16 Tekeköy, 17 Tıkalı, 18 Wadi el-Hammam, 19 Yağlak, 20 Yassıkaya (Compiled by A. Yamaç, Google Maps)

with 3 main strata. Stratum 2 yielded microliths and chipped stone tools, mostly scrapers, flakes, and blade points. The presence of bone awls implies that other materials have also been used in the manufacture of tools. The subsistence of the cave dwellers included mountain goats, deer, and wild pigs. Besides, partially burned human bones including a mandible, a skull's frontal bone, and femur fragments were found. The human bone analysis revealed that the mandible was originally a female of 18 years old (Bostancı, 1963a). Bostancı believes all the strata date back to the early Epipaleolithic Period. The microlith industry continues in the 3rd stratum. There is little distinction in the chipped stone industry between this stratum and the 2nd stratum above. This layer yielded long and thin backed blade points, which is a characteristic of the Upper Paleolithic, along with microliths (Bostancı, 1962).

2.2.3 Çarkini Cave

This cave is located northwest of the city of Antalya and 2 km southwest of the Karain Cave. Kökten opened a small test trench in the cave in 1957. Despite the lack of detailed reports on excavations, the stratigraphy has been published. The uppermost layer in this cave is Chalcolithic and overlays the Neolithic and Upper Paleolithic layers. The layer assigned to the Neolithic yielded both dark burnished ware and ware with linear designs painted in cream color. Samples of obsidian were also recovered (Kökten, 1958). Another team, led by I. Yalçınkaya, I. Kayan, and A. Minzoni-Deroche, resumed excavations at the site in 1984. The excavation of the cave and the sifting of Kökten's spoil enabled the remaining artifacts such as chipped stone debitage products and pottery to be collected more carefully. Yalçınkaya explains that Kökten had already excavated most of the cultural deposit. From the remaining artifacts, Yalçınkaya believes that the Paleolithic levels at the site fall between the Middle Paleolithic and the Epipaleolithic. The re-excavation and sifting of the soil in 1984 provided 215 chipped stone fragments (Yalçınkaya, 1995).

2.2.4 Direkli Cave

This cave is located near the village of Döngel, 38 km northwest of the city of Kahramanmaraş. K. Kökten discovered it in 1958, and a test trench was opened near the entrance the following year. The artifacts in this layer are assigned by Kökten to the end of the Upper Paleolithic. Some blade knives, drills, scrapers, and cores all made of flint were found from this layer (Kökten, 1960). This site's faunal remains are very important as well. The teeth and skeletal fragments of several Upper Paleolithic carnivores and herbivores including bear, bison, deer, wild pig, and beaver have been recovered. The second excavation phase started in 2007 in Direkli Cave by Erek (2009). Excavations continued in two separate trenches, and a "basin-like" grave was yielded at the Epipaleolithic level. The filling of the grave produced a large number of beads made of bone, stone, marine, and terrestrial shells (Fig. 2.5). No human remains have been found, however. This level's C14 date is 12.730 BP. The excavations at the same archaeological level gave rise to a baked clay figurine of a woman. During excavations, very esthetically and elaborately made bone tools were discovered. These bone tools included an eyehole needle, a ball-headed needle, a bone spear, and a spatula. The animal bone finds found within the cave mostly belonged to herbivores. Sheep, goats, antelopes, and boars are among the roughly observed species (Erek, 2009). The analysis carried out on Direkli Cave's faunal remains shows that the cave was used as a short-term logistics camp to hunt wild goats (*Capra aegagrus*) which lived during the Late Epipaleolithic Period. The fact is that the low density of the remains and the seasonal occupation data indicate that the Epipaleolithic hunter-gatherers were highly mobile.

2.2.5 Dülük Cave

This cave site is 11 km north of the city of Gaziantep (Fig. 2.6). Even though it was renown for a long time within the science world, this cave was first researched and began to be excavated in 1971 by E.Y. Bostancı. The cave is 84 m long and 24 m deep. In the 2-m-thick deposit that was excavated, bedrock was not reached. Paleolithic material was found only in one corner of the cave (Fig. 2.7). Bostancı mentions that he found examples of Upper Paleolithic blade technology. Also on the terrace near the entrance of the cave, he found a large, 18.4-cm-long biface axe (Bostancı, 1963b, 1975). The two Mithraea found within Dülük Cave that belonged to Mithra culture are dated between first century BC and third century AD. These two Mithraea found within the cave side by side share the same entrance but harbor two different cult pictures. There are no other Mithra cult areas that carry these features within all the Roman Empire. Even though being highly deformed, known from other findings gathered from other corners of the empire, the scene of Mithra kneeling on the moon bull and killing it by slicing its throat can be seen on these cult glyphs. The blood flowing from the bull's throat means a new life. Other animals in this legend like scorpion, snake, and dog can be seen despite the late period deformations. The scene is framed with the torch-bearers: Cautes and Cautopates. The top part of the glyph is shaped with an arch. Sun god Sol and moon god Luna reside at the top two corners. On the glyph at the first cult cave, Mithra's head had been shattered and a cross had

Fig. 2.5 Direkli Cave (Photograph I. Demir)

Fig. 2.6 Dülük Cave (Photograph S. Savcılı)

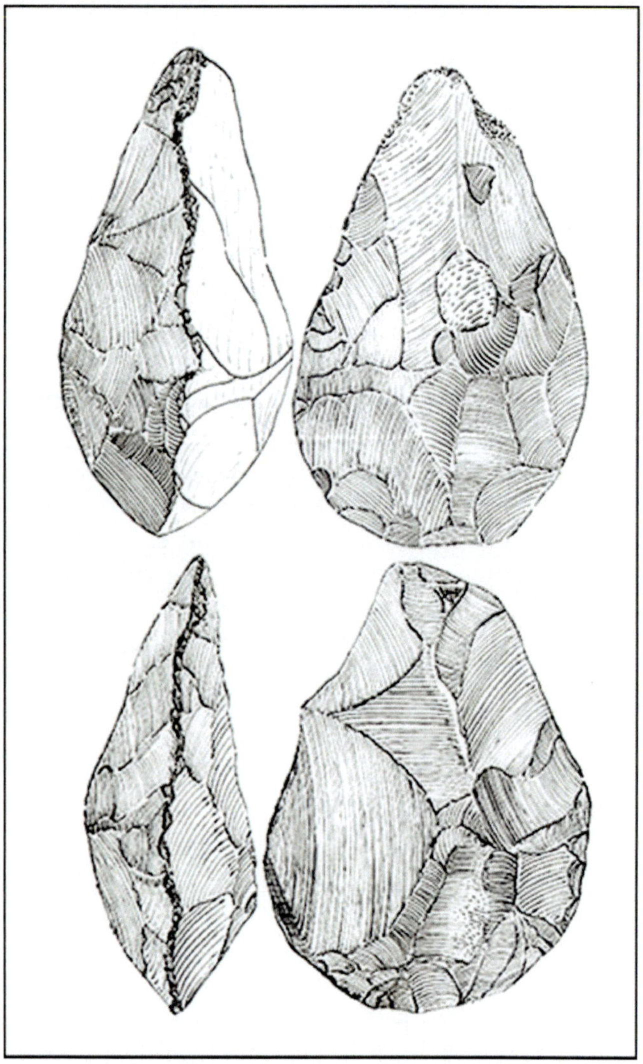

Fig. 2.7 Dülük Cave paleolithic findings from the first excavation season (Bostanci, 1963b)

been carved instead of it (Schütte-Maischatz & Winter, 2001; Winter & Schütte-Maischatz, 2004).

2.2.6 Gurma Cave

Gurma cave, which is residing west of Antalya and was excavated by Viale for the first time in 1922 (Viale, 1925–26), was re-excavated in 1946 by Kökten who extended the test trench opened by Viale and opened an adjacent 1.5 × 4 m trench. Kökten points out that the top 75 cm is Greco-Roman, that the Early Bronze Age starts 95 cm below the surface, and that virgin soil is 130 cm below the surface. Ceramic finds include wares that are burnished black on the inside and outside and wares which are burnished black on the inside and red on the outside. Kökten identifies these as Early Bronze Age wares (Kökten, 1947). The excavation of Viale yielded several triangular flint bifaces (Fig. 2.8), and one chipped stone tool was found by Kökten. Because the excavations at this site in 1922 and 1946 were less thorough in their methods of excavation and recording, like many excavations in the early twentieth century, the stratigraphic sequencing of this site could not be determined.

2.2.7 İkiağızlı Cave

The cave is located on the same hill as the Üçağızlı Cave, 12 km from Hatay's Samandag District. It was discovered during E. Güleç's 2002 survey, and she performed a brief sounding in 2008. Eight levels were identified as a result of the sounding performed in 2008. During this exploration, 305 pieces of tools and fragments, 19 cores, and 1471 pieces of remains were found. The technology that has been identified within the cave consists of the flake industry. Most of the flakes are flat, typical Middle Paleolithic tools are found within the cave, and there is a high amount of side scrapers. The Paleolithic tool density is low (Güleç, Yilmaz, Sagir, Ozer, & Baykara, 2010).

2.2.8 İncili Cave

This cave is located southwest of Antakya city, south of Çevlik Village. The cave lies in the limestone cliffs formed during the Tertiary Period, which today are directly sheltered from the sea because they are in a small bay. The cave is assumed to have been formed in the Quaternary Period by wave action. The cave's entrance is 6 meters above sea level. The cave was excavated in 1970, 1971, and 1973 under the direction of E. Bostancı, reaching a bedrock at 5 meters below the surface. The first 9 layers are post-Roman. The findings from these layers were not reported, which are totaling 274 cm in thickness. Layer 10, which is 53 cm thick, is contemporaneous with the settlement of the ancient city of Selevkia Pierra. From this light-colored calcareous layer, large quantities of pottery and broken glass were recovered. The underlying layer was a 40-cm-thick grayish-white stratum that yielded many fragmented pieces of human bones. Layer 12 consisted entirely of sand formed at a time when sea level was lower. Bostancı dates Layer 12 to the Upper Paleolithic. Although no chipped stone tools have been found, the human skeleton and the pierced shells, which are at least 50,000 years old, were probably used as decorative ornaments (Bostancı, 1973b).

2.2.9 İnönü Caves

A team led by Şevket Aziz Kansu made an excavation by the name of Turkish Historical Society in Eskişehir İnönü caves in 1938 (Kansu, 1939). This is the very first Turkish archaeological survey and excavation. There are many caves and rock shelters of different sizes in the area (Fig. 2.9). Kansu and his team had excavated largest two of these caves which they named A and B. A typical stratigraphy was detected, and pottery fragments belonged to Chalcolithic and

Fig. 2.8 Paleolithic flint tools from 1922 Gurma Cave excavations (Viale, 1925–26)

Fig. 2.9 İnönü Caves (Photograph N. Çam)

Bronze Ages were found but no findings belonged to Paleolithic Age were encountered during the excavation made in cave A. A small amount of Bronze Age pottery was found during the test excavation made in cave B and its side branches. Caves and rock shelters in the area were also researched during these excavations. These are Eserönü Cave to the east of İnönü, Suluin, Koyunini, and Çakmakini caves of Kandilli Village to the south of İnönü, three caves on the İnönü–Kütahya road, Karahasan Rock Shelter in Porsuk Suyu Valley, and the two caves found in the foothills of the Kırgız Mountain to the east of Arapören Village. İnler and Katrancı Village caves found approximately 70 km south and west of Polatlı, Kızılhisar, and Demirözü caves found in the valley which Demirözü creek flows through were examined and found poor in terms of prehistoric material. Kirmir Suyu caves in Güdül, Ankara, were examined the same year and were deduced to belong in classical periods. Kale Cave, which is located in a hill called Kale to the south of Şereflikoçhisar, was also examined.

2.2.10 Kanal Cave

This cave site is located southwest of Antakya city and very close to Mağaracık Village (Fig. 2.10). It is about 500 m from the Merdivenli Cave and 300 m from the Mediterranean. The cave's entrance is 20 meters above sea level. The site was found in 1966, and excavations were conducted in 1966, 1967, and 1969 by Enver Yaşar Bostancı. Several different sand strata are reported to be found among the Upper Paleolithic terra rosa iron oxide layers. The researchers believe this indicates that in the Upper Paleolithic, this region's climate was dry. The Mediterranean, which in the Riss-Würm Ice Ages was 15 m higher than it is today, was probably much closer than it is today to this cave. Layer VI, the lowest excavated layer, is particularly difficult. Minzoni-Deroche reports that the cultural material filling at this site consists of mixed deposits that have been hardened by water erosion and sloping sediment movement (Minzoni-Deroche, 1993). This explains why the excavated strata report lacks clarity. At this site, the first test trench dug revealed the presence of an Upper Paleolithic. Excavating further exposed Middle Paleolithic remains. The Upper Paleolithic layers yielded end scrapers on blades, rounded scrapers, biface end scrapers, steep scrapers, burins on flakes, and blades (Bostancı, 1967a).

Bostancı reports that there is a total of 20,000 artifacts and debitage products. The excavation also yielded a few human remains including a mandible molar in the Upper Paleolithic layers and a canine tooth from a 9-year-old boy's maxilla in the Middle Paleolithic strata. A lot of fossilized mammal bones were also collected here, although not well described. Minzoni-Deroche reports that in the surface survey she conducted after the excavation, she found 1468 tools and tool fragments at this site. The findings included rounded end scrapers; side scrapers; burins on blades; borers, etc., assigned to Upper Paleolithic. Middle Paleolithic tools are not mentioned (Minzoni-Deroche, 1993).

2.2.11 Kapalıin Cave

This cave site is located 1.7 km northwest of the village of Senirce, 15 km north of Isparta city. Other caves and rock shelters on the same cliff are collectively called the Bozaönü Caves (Fig. 2.11). The entrance of the Kapalıin Cave faces

Fig. 2.10 Kanal Cave (Photograph E. Soner)

south. The cave is 21 m deep and 21.5 m wide, consisting of two different chambers. The larger chamber, closer to the entrance, has a high conical ceiling, whereas the smaller chamber has a low ceiling and there are several broken stalactites on the ceiling. The cultural fill over the cave's floor seems to be quite thick. In the Byzantine Period, both the interior and the exterior of the narrow entrance had been neatly reshaped. While outside the entrance a wooden porch-like structure was built, a stepped arch was carved on the ceiling of the cave entrance. Excavations at the site were conducted in 1944 by Ş. A. Kansu and 1951 and 1952 by M. Şenyürek. Two trenches were opened during the excavations of 1944 (Kansu, 1944).

These test trenches were expanded by later excavations, and 4 new trenches were opened. The stratigraphy is as follows: The top two layers (layers I and II), with a total depth of 40 cm, consist of manure and chips of calcareous stone. The layer below is 45 cm thick and dates back to the Roman and Byzantine periods. Layer IV is a reddish ash layer with a thickness of 35 cm and includes fragments of worked calcareous and burned bone. The burned bone fragments continue in layer V. This layer, dating back to an unspecified "later period," yielded a large amount of pottery and tiles. Layer VI is a sandy layer with fragments of worked limestone. These top six layers represent this site's historical periods. In the level below, a large decrease in the amount of pottery can be seen. Small pebbles, flint and bone tools, and fossilized animal bones assigned to the Upper Paleolithic were found in the 75-cm-thick layer VII. The dark-colored layer below is full of chipped stone tools. Interestingly, ceramics are reported to reappear in this 85-cm-thick layer. Therefore, this site's stratigraphy needs to be researched further. The next layer, layer IX, is a thin yellowish layer that lies above the ashy layer X, which is the final material cultural level. There was no pottery in level IX or level X. Kansu reports that the bedrock was reached 3.5 m below the surface, while Senyürek reports that the earliest layer of cultural deposit was layer X. Kansu reports that the first Upper Paleolithic finds were discovered in 1944. There are many discrepancies in stratigraphy that require clarification.

Fig. 2.11 Kapalıin Cave (Photograph M. Çamlıalan)

Although the stratigraphic layers are presented in detail, there is limited information on the microlith chipped stone tools. Microburins found during the excavation were lunates, triangles, and other geometric microtools. Since all the tools described and photographed are Epipaleolithic, it seems reasonable to classify the site as Epipaleolithic until further research.

2.2.12 Keçiler Cave

This cave is located 3 km north of Büyük and Küçük Pirun villages, 6–7 km northeast of Adıyaman city. There are two separate entrances to the cave which is 20 m long. Hours refer to this cave as "Palanlı Cave" (Hours et al., 1994). Anati, who arrived in the region to analyze the cave paintings of the nearby rock shelter Palanlı–Pirun, discovered the site in 1968 (Anati, 1968). Bostancı later surveyed the cave in 1970 (Bostancı, 1971). Anati reports that the rock face contains at least 45 engraved figures, particularly on the eastern wall. Anati, who has studied all the figures, believes they have been drawn in four phases. Especially the first and oldest phase is deeply engraved. It depicts well-pictured goats, ranging in height from 60 to 80 cm (Fig. 2.12). There are also two schematic human figures in this phase. Anati dates this phase to the Epipaleolithic Period except for a few drawings that she believes are earlier, while Bostancı assigns it to the Upper Paleolithic. Anati believes that the drawings in this phase closely resemble the paintings in Italy's Romanelli, Levanzo and Adduara caves, as well as Jordan's Negev Desert and Kilwa Cave art paintings. On the other hand, the paintings produced in the second and third phases were chiseled. Anati, who believes they are similar to the examples of cave art from the Kumbucağı Rock Shelter and the Gevaruk Valley, assigns them to the Neolithic Period, while Bostancı believes that they also have some Epipaleolithic qualities. Some drawings were superimposed over the earlier engravings in Phase Four. While Anati dates these drawings to the Bronze Age, they are assigned to modern times by Bostancı (Anati, 1968; Bostancı, 1973a). Anati compares the rock art in the second and third phases to the wall paintings in Beldibi and the Gevaruk Valley wall paintings, concluding that it is of the Neolithic Period (Anati, 1968). On the other hand, Bostancı assigns the rock art to the Epipaleolithic and Early Neolithic as it is stylized (Bostancı, 1973a). However, Mellaart believes that the animal figures and other depictions of the rock art in this cave resemble Halaf painted pottery and concludes that the paintings should be assigned to the Chalcolithic Period instead (Mellaart, 1975). In ASPRO, those cave paintings are dated to 9600–8000 BP (Hours et al., 1994).

2.2.13 Kızılin Cave

Not to be mixed with Burdur's Kızılin Cave mentioned in Chap. 5, this cave site is located northwest of Antalya city, southwest of the Karain Cave. Kızılin Cave, like the caves of Karain, Çarkini, and Öküzini, is on the northern fringes of Mount Katran, facing the travertine formation. The cave is 410 m above sea level and overlooks a plain approximately 100 m above the level of the plain. The dimensions of the cave are 20 m east–west and 14 m north–south. Today, the cave's entrance is only 1 m wide and it is presumed that the entry was wider in the Pleistocene. Although there is a flat terrace by the front entrance, the cave's floor runs on a steep slope. It is known that Kökten excavated at this cave site but his findings have not been published. Işın Yalçınkaya, Ilhan

Fig. 2.12 Keçiler Cave (Photograph E. Soner)

Kayan, and Angela Minzoni-Deroche surveyed the area later in 1984. During their survey in 1984, the entrance to the cave yielded 450 flintstone flakes and tool fragments. A total of 309 of these were blades and flakes, some of which were retouched. Also, some end scrapers and cores were found. The most frequently found tools were blades and end scrapers which were notched and slightly retouched. A rich faunal assemblage was recovered from this cave. The researchers suggest that the cave must have been inhabited in post-Glacial, Upper Paleolithic times and that although common Upper Paleolithic chipped stone artifacts have not been found, such tools are expected to be found in future excavations (Minzoni-Deroche, 1988, 1993; Yalçınkaya, 1986).

2.2.14 Kurbanağa Cave

This cave is located south of Kars city, southwest of Camuşlu Village. The cave's depth is 11.50 m, and it has a 5.50 m interior width while its entrance is 12.50 m wide. The cave art on the outer walls and the ceiling of the cave were made by chiseling, which is a technique foreign to this area. The panels depict goats, some with arrows shot into them; traps made of nets; wooden stakes; and braided lassos. These wall paintings unmistakeably date back to the Paleolithic. Kılıç Kökten discovered the cave in 1969, opening a 3.5 × 8 m test trench just outside the cave the same year. Although it is indicated that the upper levels of the site date to the Early Bronze Age, it is unfortunate that detailed information on the strata and the finds have not been published. Upper and Middle Paleolithic tools were found together in the layer below (Kökten, 1975).

2.2.15 Merdivenli Cave

This cave site resides southwest of the city of Antakya, southwest of Mağaracık Village and 4 km northwest of Samandağ town. The cave, now 36–39 m above sea level, is about 23 m deep, 6 m wide, and 2 m high. The cave's entrance faces south. Two seasons of excavation were carried out in 1956 and 1957 after the discovery of the cave in 1956. Under the direction of M. Şenyürek and E. Bostancı, six test trenches were excavated. A total of five layers were exposed according to the reports of the researchers. In this excavation, bedrock was not reached at the final depth of 571 cm. Some oil lamps and some pottery fragments were found at the first layer, which dates back to the Roman Period when the area was settled by the inhabitants of the city of Selevkia. Points, blades, end scrapers, nosed scrapers, flake scrapers, steep faced scrapers, and borers assigned to the Upper Paleolithic were found in the second and third layers. Some bone tools, including a hatchet, an awl, and the bottom portion of a needle were also found in these layers. There is a thin layer between layer III and layer IV which contains stones that must have fallen off the cave's ceiling. There were no artifacts in this layer. The fourth layer, partially hardened by a calcareous deposit, included flint and bone fragments. These layers vary in thickness in each trench. Chipped stone tools of the Levallois–Moustérien type have been found in the last two layers. These include triangular points, laurel leaf points, side scrapers, perforators, and concave disk-shaped scrapers. In layer IV, there is a notable advance in the chipped stone tool industry when compared to the layer V. Very comprehensive site reports on the chipped stone tools of this Paleolithic cave site were published by Şenyürek and Bostancı. There were many fewer bone tools in the fourth and fifth layers. There is a layer of large stones above the sandy strata below layer V. The researchers suggest that there was a great time gap in the Middle Paleolithic between the deposition of these sandy strata and the habitation. Four human molars and mammal bones, including lion, hippopotamus, deer, bison, as well as many other fossils from a variety of carnivorous animals and rodents, were found among the layers assigned to the Middle Paleolithic.

2.2.16 Tekeköy Caves

These rock shelters are about 14 km southeast of Samsun City, south of Tekeköy Village (Fig. 2.13). The survey and excavation of Tekeköy caves were carried out in 1940 under the direction of Nimet Özgüç, Tahsin Özgüç, and Kılıç Kökten, and this excavation is special because it was the very first cave excavation in the Black Sea Region (Kökten, Ozguç, & Ozguç, 1945). Kökten's stratigraphy does not correlate with that of Kansu, who visited the site in 1941 (Kansu, 1944). Kökten found many chipped stone tools on the valley terraces as well as in the rock shelters in his survey of the neighboring caves and rock shelters in 1940 (Kökten, 1941, 1944). Retouched microlith tools are reported to be abundant in rock shelter "A" within the layers II and III, as defined by both Kansu and Kökten. Kansu places the tools, which are resembling northwestern European microlith industries tools, in three different categories: microburins, microscrapers, and other microlith tools (Kansu, 1944). Even though the excavations provided a large number of animal bones and teeth including buffalo and horse, no faunal research was carried out. Kökten reports that the tools are made of a variety of materials, while the smaller of the two Moustérien tools he found on the terrace of the valley is a diorite scraper, and the larger one is a point made from basalt.

Fig. 2.13 Tekeköy Caves (Photograph A. E. Keskin)

2.2.17 Tıkalı Cave

This cave site is located in the village of Mağaracık, southwest of the city of Antakya. It was discovered during their survey and excavations in the Samandag region in 1958 by M. Şenyürek and E Bostancı. Two test trenches were opened the same year, a small one at the cave entrance and a larger 4.7 × 4 m one west of the small trench. The excavations continued with new trenches the following year. According to the excavators, Middle Paleolithic tools appeared between 145 and 180 cm below the surface. Roman drainage pipes found just below the uppermost layer, which consists of the rubble that fills the cave, indicate that the rubble must have fallen off the cave ceiling in relatively recent times. Şenyürek reports that the thickness of the Paleolithic layer below is 135–145 cm. Hardened soil with chipped stone tools and fossilized mammal bones were at the bottom of this layer. Excavations were stopped before bedrock was reached because it was too hard to excavate this layer. Excavations in the Tıkalı Cave unearthed Middle Paleolithic chipped stone tools that are among the most carefully examined Anatolian Paleolithic finds. Although some of the flakes were made from hard sandstone, most of the tools and debitage products were made from silica. The excavators believe that the differentiating patinas are the result of flintstone pebble usage in the making of tools instead of flint nodules. Flakes and blades are prominent components of this industry, but points, side scrapers, end scrapers, burins, borers, knives, and choppers were also found (Senyürek, 1958, 1959).

2.2.18 Wadi el-Hammam Cave

This site is located 1.7 km southeast of Reyhanlı town, southeast of Antakya city. The cave resides 500 m southwest of the Tell el Cüdeyde/Judaidah site and is one of many caves in the Amuq Plain's limestone outcrops (Fig. 2.14). The cave was discovered in 1932 by M. Claude Prost and Terence Patrick O'Brien directed an excavation in the cave. The excavation, which is Anatolia's second earliest cave excavation after the Gurma Cave excavation led by Viale, uncovered an area of 48 m^2, which was excavated 2 m below the surface. In addition to small finds, the floors and surfaces inside the cave suggest that there was habitation in the cave. There were four layers found. Layer 4 has been assigned to the Neolithic Period, which is 166 cm beneath the surface (O'Brien, 1933). There is little difference in the material found on different floors within the cave. Amuq A handmade, thick pasted, simple kitchenware, and dark burnished ware with black, gray, brown, gray-red surface colors were found in the cave. Slipped ware was found in the lower levels. Amuq B impressed decorated ware was found in the uppermost levels. Points, spearheads, scrapers, obsidian retouched blades have been found. The findings do not provide specific information about the chipped stone industry. Green beads made of calcite have been found. Skeleton of a newborn baby was excavated along with an adult skeleton. This site was placed in both the fifth and sixth phases in ASPRO (Amuq A and B) (Hours et al., 1994).

Fig. 2.14 Photograph showing Wadi el-Hammam from the north. Wadi el-Hammam Cave is on the left side of the photograph (O'Brien, 1933)

2.2.19 Yağlak Cave

This cave site is located 38 km northwest of Kahramanmaraş, 800 m east of Döngel Village. The surrounding area is karstic, and around Yağlak Cave several caves lie in steep cliffs overlooking the deep valleys (Fig. 2.15). Soon after discovering this site, Kökten opened a test trench within the cave in 1959. The uppermost layer of this small trench opened entirely by Kökten's efforts is a 110-cm-thick black, humusy, loose layer which provided a scatter of mixed artifacts. The second layer, which Kökten assigns to the Neolithic, is approximately 100 cm thick and further beneath is a Paleolithic layer with a thickness of 66 cm, rich in microlithic tools made from silica (probably a transitional phase between the Late Upper Paleolithic–Epipaleolithic periods) (Kökten, 1960). It is not reported whether bedrock was reached. While Kökten reports that he had done more research in the area in 1960–61, his findings were not published. There seems to be no intermediary level between the Upper Paleolithic and the Neolithic periods in this cave, unlike the caves of Karain and Öküzini.

2.2.20 Yassıkaya Cave

The site is situated about 20 km east of Karadeniz Ereğlisi, on a valley's southern ridge, near Ovaköy. Both sides of the valley are rocky, while trees cover the slopes and flat areas. The site is located on the valley's southern edge, on a rocky plateau that stretches across a large area. The cultural deposit was defined to be in a natural cave in the middle of the rocky area and on the adjacent platform. The cave entrance is facing the valley in the north–south direction. Excavations in this settlement were conducted under the scientific advisory of Turan Efe from Istanbul University in 2000. The color of the bottom layer is dark gray. The top layer is yellow and mixed with broken stones. Upper layers are fully dated as Early Bronze Age (Fig. 2.16). Burnt sherds and mudbrick pieces with pole stains have been found in this fill. There was also a similar cultural fill with two layers on the platform. There is a clear difference between the layers of culture. There are three groups of pottery: plain courseware, red slipware, and dark rim ware. Deep bowls, spouted pitchers, and inverted rimed or wide-necked pots are the most typical forms. Relief bands characterize the decorations on the pots. There are knobs on some examples close to both sides of the handles. Some potsherds also display incised decoration. Spindle whorls are mostly half sphere-shaped, but one or two samples are conical and few of them are decorated. Loom weights are shaped slightly flat, and they have rounded heads. There were also some flint blades and basalt grinding stones found. It is one of the few prehistoric sites in the region of the Black Sea (Efe, 2004).

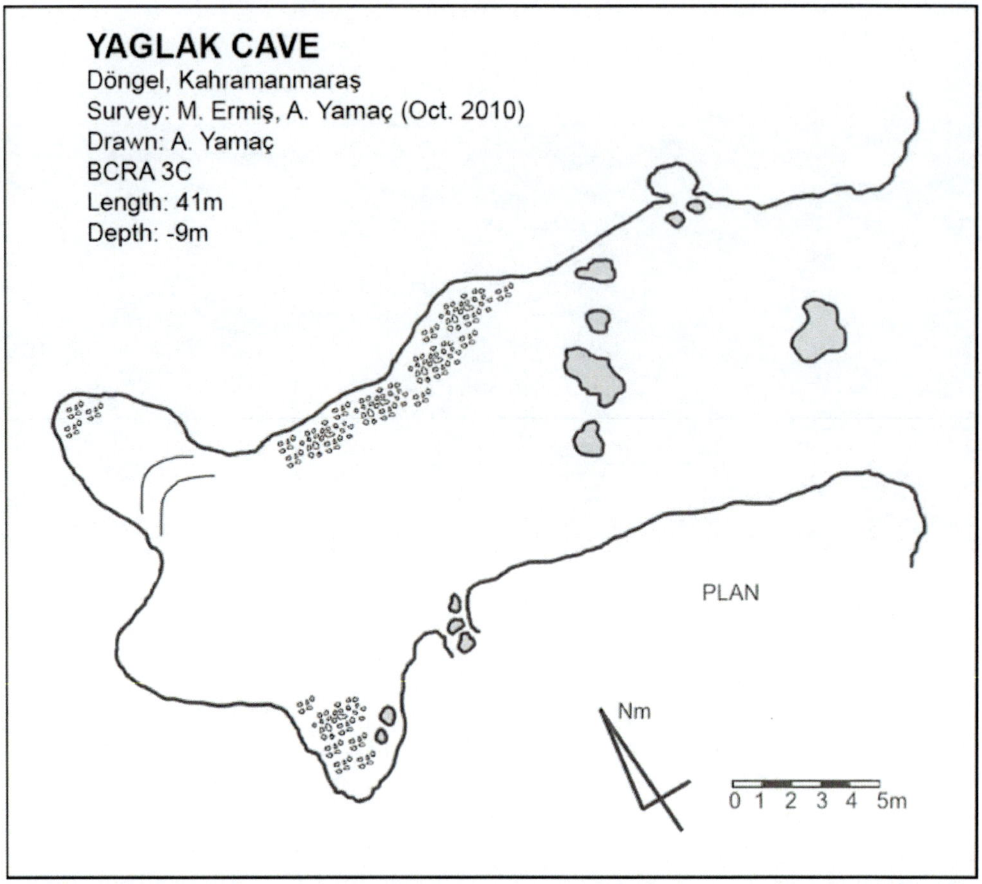

Fig. 2.15 Plan of Yağlak Cave (drawn by A. Yamaç)

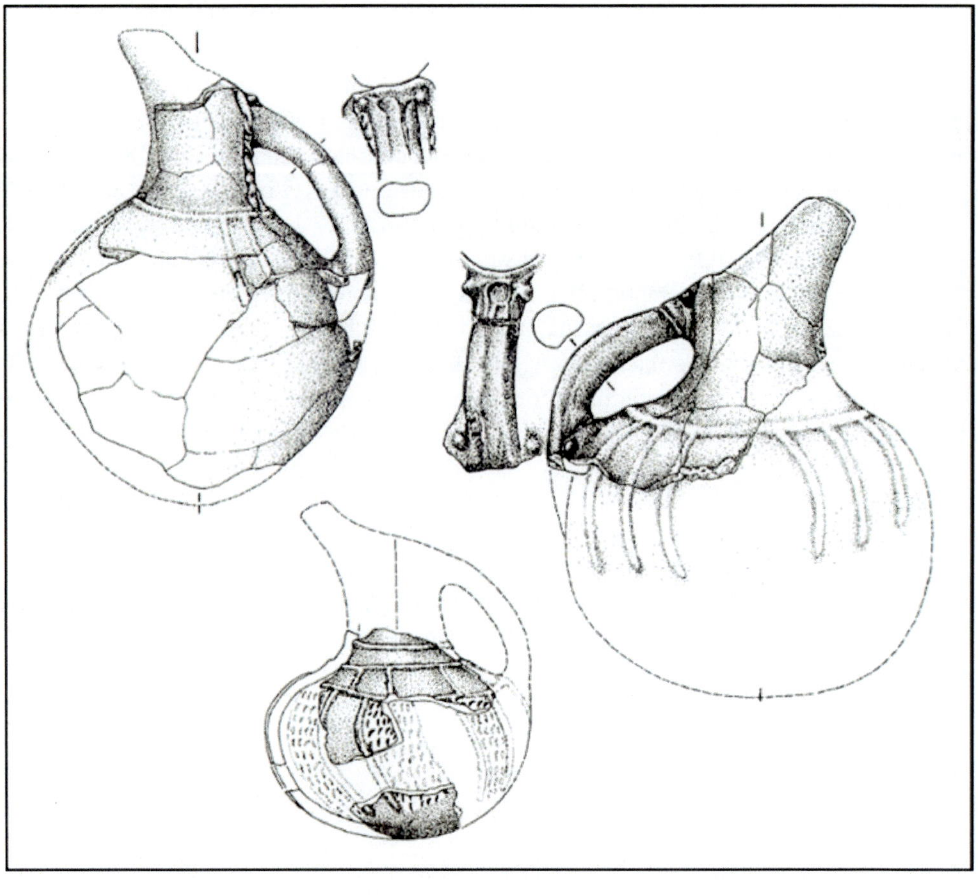

Fig. 2.16 Early Bronze Age pottery from Yassıkaya Cave (Efe, 2004)

References

Anati, E. (1968). Anatolia's Earliest Art. *Archaeology, 21,* 22–33.

Arsebük, G. (1998). *Yarımburgaz Mağarası. TUBA-AR, 1,* 9–25.

Bostancı, E. (1959). *Research on the Mediterranean Coast of Anatolia; a new palaeolithic site at Beldibi near Antalya; Preliminary Report.* Retrieved from Ankara.

Bostancı, E. (1962). A New Upper Palaeolithic and Mesolithic Facies at Belbaşi Rock-Shelter on the Mediterranean Coast of Anatolia. *Belleten, 26*(252–272), 276–278.

Bostancı, E. (1963a). *Fossil remains of Upper Palaeolithic and Mesolithic Man in Beldibi and Belbaşi Rock-Shelters on the Mediterranean Coast of Anatolia.* Paper presented at the Bericht über die 8. Tagung der Deutschen Gesellschaft für Anthropologie, Köln.

Bostancı, E. (1963b). Researches in South-East Anatolia. The Chellean and Acheulean Industry of Dülük and Kartal. *Anadolu, 6,* 111–141.

Bostancı, E. (1966). Important artistic objects from the Beldibi excavations. *Antropoloji, 1,* 25–31.

Bostancı, E. (1967a). Mağracık Çevresinde Yapılan 1966 Yaz Mevsimi Kazıları ve Yeni Buluntular. *Antropoloji, 3,* 19–45.

Bostancı, E. (1967b). The Mesolithic of Beldibi and Belbaşi and the Relation with the Other finding in Anatolia. *Antropoloji, 3,* 91–142.

Bostancı, E. (1971). A research on the solutreen and Adıyamaniyen cultures surrounding of Adıyaman. *Antropoloji, 5,* 45–82.

Bostancı, E. (1973a). Güney-Doğu Anadolu'da Adıyaman Çevresinde Yapılan Paleoantropolojik Prehistorik ve Quaterner Problemleri Üzerinde Bir Araştırma. *Antropoloji, 6,* 145.

Bostancı, E. (1973b). Homo Sapiens Çevlikensis in the Canal and the Big Caves near Samandağ of the Province of Antalya on the Mediterranean Coast of Anatolia. *Antropoloji, 6,* 29–56.

Bostancı, E. (1975). İnsan Evriminde Okucu Kültürü Anadoluda Dülükiyen Alt Taş Kültüründe; Keşfedilen en Eski Acheuleen Devre Ait Tipik Bir Ok ucu. *Antropoloji, 9,* 15–45.

Dinçer, B. (2016). The Lower Paleolithic in Turkey: Anatolia and Hominin dispersals Out of Africa. In K. Harvati & M. Roksandic (Eds.), *Paleoanthropology of the Balkans and Anatolia* (pp. 213–228). Springer.

Efe, T. (2004). Yassıkaya, an Early Bronze Age Site near Heraclea Pontica (Kdz Ereğli) on the Black Sea Coast. In B. Hänsel (Ed.), *Festschrift für Němejcová Pavùková* (pp. 27–38). Berlin.

Erek, C. (2009). *2007 Yılı Direkli Mağarası Kazıları.* Paper presented at the 30. Kazı Sonuçları Toplantısı, Ankara.

Güleç, E., Özer, I., Sağır, M., Baykara, I., & Şahin, S. (2013). *2011 Yılı Gaziantep ve Hatay İlleri Yüzey Araştırması.* Paper presented at the 30. Araştırma Sonuçları Toplantısı, Ankara.

Güleç, E., Yilmaz, H., Sagir, M., Ozer, I., & Baykara, I. (2010). *2008 Yılı Hatay ve Eskişehir İlleri Yüzey Araştırması.* Paper presented at the 27. Araştırma Sonuçları Toplantısı, Ankara.

Harmankaya, S. (1997). Türkiye Paleolitik araştırmaları üzerine bir değerlendirme. In S. Harmankaya & O. Tanındı (Eds.), *Türkiye Arkeolojik Yerleşmeleri I: Paleolitik-Epipaleolitik* (pp. 7–19). Istanbul: Ege Yayınları.

Harmankaya, S., & Tanındı, O. (1997). *Türkiye Arkeolojik Yerleşmeleri I: Paleolitik-Epipaleolitik.* İstanbul: Ege Yayınları.

Harmankaya, S., Tanındı, O., & Özbaşaran, M. (1997). *Türkiye Arkeolojik Yerleşmeleri 2: Neolitik.* İstanbul: Ege Yayınları.

Hours, F., Aurenche, O., Cauvin, J., Cauvin, M., Copeland, L., & Sanlaville, P. (1994). *Atlas des Sites du Proche Orient (14000–5700 BP).* Paris.

Kansu, Ş. (1939). Türk Tarih Kurumu Prehistorik Araştırmaları. *Belleten, 3*(9), 93–97.

Kansu, Ş. (1944). Anadolu'da Mezolitik Kültür Buluntuları. *Dil Tarih ve Coğrafya Fakültesi Dergisi, 2*(5), 673–682.

Kökten, I. (1941). Samsun Vilayeti Tekeköy Civarında Prehistorik Araştırmalar. *Ulkü, 98*(17), 121–124.

Kökten, I. (1944). Die Bedeutung der Umbegung von Bayburt İnnerhalb der Vorgeschichte nordost-Anatoliens. *Dil Tarih ve Coğrafya Fakültesi Dergisi, 3*(5), 487–505.

Kökten, I. (1947). Bazı Prehistorik İstasyonlar Hakkında Yeni Gözlemler. *Dil Tarih ve Coğrafya Fakültesi Dergisi, 5,* 223–236.

Kökten, I. (1952). Anadolu'da Prehistorik Yerleşme Yerlerinin Dağılışı Üzerine Bir Araştırma. *Dil Tarih ve Coğrafya Fakültesi Dergisi, 10* (3-4), 167–207.

Kökten, I. (1958). Tarsus-Antalya Arası Sahil Şeridi Üzerinde ve Antalya Bölgesinde Yapılan Tarih Öncesi Araştırmaları. *Türk Arkeoloji Dergisi, 8*(2), 10–16.

Kökten, I. (1960). Anadolu-Maraş Vilayetinde Tarihten Dip Tarihe Gidiş. *Türk Arkeoloji Dergisi, 10*(1), 42–52.

Kökten, I. (1975). *Kars Çevresinde Dip Tarih Araştırmaları ve Yazılıkaya Resimleri.* Ankara: Paper presented at the Atatürk Konferansları V.

Kökten, I., Ozguç, N., & Ozguç, T. (1945). 1940 ve 1941 Yılında Türk Tarih Kurumu Adına Yapılan Samsun Bölgesi Kazıları Hakkında İlk Kısa Rapor. *Belleten, 9*(35), 361–400.

Mellaart, J. (1975). *The neolithic of the near east.* London: Thames and Hudson.

Minzoni-Deroche, A. (1988). Le Paléolithique Moyen du Taurus. *Paléorient, 14*(2), 154–158.

Minzoni-Deroche, A. (1993). *Middle and Upper Palaeolithic in the Taurus-Zagros Region. The Paleolithic Prehistory of the Zagros-Taurus* (pp. 147–158). Pennsylvania: University of Pennsylvania.

Moretti, G. (1923–1924a). In Daghindà Quogia In, La Grande Caverna nella Montagna delle Caverne. *Annuario della Reale Scuola Archeologica di Atene, 6–7,* 509–546.

Moretti, G. (1923–1924b). Le grotte sacre di luvadja. *Annuario della Reale Scuola Archeologica di Atene, 6–7,* 547–554.

O'Brien, T. (1933). A Chalcolithic Cave Site in North Syria. *Man, 33,* 173–178.

Otte, M. (1998). Turkey as a key. In A. et al. (Ed.), *Neandertals and modern humans in Western Asia.* New York: Plenum Press.

Özdoğan, M., & Karul, N. (2002). *Şanlıurfa; Birecik-Suruç Arkeolojik Envanter Raporu.* Retrieved from Ankara.

Schütte-Maischatz, A., & Winter, E. (2001). Die Mithräen von Doliche. Überlegungen zu den ersten Kultstätten der Mithras-Mysterien in der Kommagene. *Topoi, 11,* 149–173.

Senyürek, M. (1958). New researches in the Hatay Province. *Belleten, 22*(87), 443.

Senyürek, M. (1959). Tıkalı Mağaranın Paleolitik Endüstrisine Dair Bir Not. *Belleten, 23,* 10–26.

Shea, J. (2013). *Stone tools in the Paleolithic and Neolithic Near East.* Cambridge: Cambridge University Press.

Taşkıran, H. (2018). *Prehistorik Arkeoloji ve Mağaralar. Mavi Gezegen, 24,* 62–68.

Taşkıran, H., & Kartal, M. (2004). *2001 Yılı Karkamış Baraj Gölü Alanı Paleolitik Dönem Yüzey Araştırması.* Retrieved from Ankara.

Viale, V. (1925–26). Relazione sull'attiva della Missione archeologica di Adalia nell'anno 1922. *Annuario della Reale Scuola Archeologica di Atene, 8–9,* 357–392.

Winter, E., & Schütte-Maischatz, A. (2004). *Doliche-Eine kommagenische Stadt und ihre Götter: Mithras und Jupiter Dolichenus*. Bonn: Habelt.

Yalçınkaya, I. (1986). *Batı Toroslarda Paleolitik Çağ Yüzey Araştırmaları*. Paper presented at the 3. Araştırma Sonuçları Toplantısı, Ankara.

Yalçınkaya, I. (1995). *Anadolu İskan Tarihinde Katran Dağı*. Paper presented at the 1994 Yılı Anadolu Medeniyetleri Müzesi Konferansları, Ankara.

Yalçınkaya, I., Kösem, B., Özçelik, K., Erek, C., & Kartal, M. (2000). *1998 Yılı Yüzey Araştırmaları ve Hatay Paleolitiği*. Paper presented at the 17. Araştırma Sonuçları Toplantısı 1999, Ankara.

3 Natural Sciences in the Caves of Turkey

Abstract

Karstic lands cover the %40 of the total area of Turkey. Examination of these areas situated in the Alpine–Himalayan Orogenic Belt plays a key role in shedding a light on the tectonic evolution of this huge area that elongates from Europe to Asia, besides the tectonic evolution of Turkey. The research of the karstic lands exerts a great importance for the urban policies within the cities and villages settled in these geological regions, in order to require rigorous approaches to the engineering problems during the stages such as road and freshwater system construction or wastewater treatment. The most well-known karstic formations and caves exert a great impact on the development of landscapes as underground drains, since caves have been investigated in the context of geomorphological setting, lithology and geologic structure, groundwater chemistry and hydrology, aquifer type as well as regional tectonics, climate research, biology, microbiology, and in-cave environmental conditions. The historical background of the particular research for each selected caves from Turkey is given in Chap. 4 of this book. As comprehensively treated in Chap. 2, the state of the art in cave archaeology was excluded from the contents of this chapter.

3.1 Introduction

Karstic lands that cover the %40 of the total area of the country are divided into six regions in Turkey: Taurus Mountain, Thrace and Black Sea Mountains, Western Anatolia, Central Anatolia, Eastern Anatolia, and Southeastern Anatolia (Chap. 4). First two of the regions which carry traces of Turkey's paleotectonic evolution that started in the Permian–Triassic period and continued in Mid-Miocene, and traces of the Alpine Orogenesis that belongs to Oligocene–Miocene period, present most of the karstic areas in Turkey. Conditions shaped through processes occurred like the formation of the Pontids and the Anatolid-Torid block in paleotectonic period, the collision of the Anatolid-Torid Platform with the Pontid island arc by which N–S crustal shortening and in crustal deformations of the Anatolid-Torid block began during Late Paleocene–Early Eocene, the uplift of the Anatolids during Upper Eocene–Lower Miocene period due to the ongoing compression in N–S direction, the closure of the southern branch of the Neo-Tethys in Late Cretaceous, and finally the collision of Arabian and Eurasian Plates, set the stage for the neotectonic period that contains recent earthquakes (Şaroğlu & Yılmaz, 1986). There is a suit of well-detailed publications on these concepts (such as Okay, 2008; Şengör & Yılmaz, 1981), which are of the mechanism that lightly referred herein to make an emphasis on the importance of the researches conducted on the areas that form karstic landscapes of Turkey. Examination of these areas situated in the Alpine–Himalayan Orogenic Belt plays a key role in shedding a light on the tectonic evolution of this huge area that elongates from Europe to Asia, besides the tectonic evolution of Turkey.

From north to south, the karstic topography in Turkey harbors many villages, towns, and city centers, which is another factor that increases the importance of karstic research. There are many highland villages found at up to 1500 m of altitude on the Taurus Mountains that elongates along the shore of the Mediterranean in the south of Turkey and on the mountains that elongate on the shores of the Black Sea in the north. Moreover, the most important touristic cities and ports of Turkey are located on Mediterranean shores on the south of the Taurus Mountains from east to west. Similarly, there are many villages and cities on the Obruk Plateau (the Konya Closed Basin (KCB), Central Anatolia), which is another important karstic area. All these circumstances compel to elucidate the surface and subsurface geomorphology of the karst within these regions and the factors that formed the features of it, to maintain the durability of these settlements. The extent and intensity of

A. Yamaç et al., *Caves and Karst of Turkey—Vol. 1*, Cave and Karst Systems of the World,
https://doi.org/10.1007/978-3-030-65501-3_3

karstification in Turkey are driven by a hierarchy of influences with a differentiating order in each region, such as climate parameters (the type and amount of precipitation, temperature, etc.), the spatial pattern of the features of tectonism and the stratigraphic/lithologic features of the rock, and hence, the features of the landscape dramatically change within short distances. Through regions, the unique complex reliefs as combinations of badlands, deep valleys, and the other karstic surface features in addition to large cavities and underground water systems emerge engineering problems during the stages of urban planning with all the services including road and freshwater system construction as well as wastewater treatment. The situation requires the technical examinations of karstic areas such as the evaluation of soil and bedrock stability, as well as the permeability, solubility, and aquifer yield assessments. In this sense, beyond searching for answers to questions of science that scale in billions of years, karstic researches have critical importance on the determination of short- and long-term governmental policies of urbanization.

The most well-known karstic formations and caves exert a great impact on the development of landscapes as underground drains. Harboring over 2000 known caves, Turkey is thought to have over 20,000 caves dispersed in karstic areas. For this reason, there are many studies conducted on karst hydrology, geology, and geomorphology in Turkey; qualities of these areas are studied with frequent and detailed geochemistry and geophysics studies, and there is an increase in the numbers of cave studies in recent years (Günay, Güner, & Törk, 2015). Caves have been investigated in the context of geomorphological setting, lithology and geologic structure, groundwater chemistry and hydrology, aquifer type as well as regional tectonics, climate research, biology, microbiology, and in-cave environmental conditions. The historical background of the particular research for each selected caves from Turkey is given in Chap. 5 of this book. As comprehensively treated in Chap. 2, the state of the art in cave archaeology was excluded from the contents of this chapter. Additionally, as it is given in Chap. 2, the concepts and region-specific processes related to geology, hydrology, and geomorphology of Turkey were not explained in this chapter.

3.2 The Research with Perspectives of Cave Geomorphology, Geology, and Hydrology

The oldest study that examines a cave with referring to its structural features, also the first known publication about cave research in Turkey, is Abdullah Bey's study conducted in Yarım Burgaz Cave in 1969 (see Chap. 2). Since then, there have been many studies on geological, geomorphological, geochemical, and hydrogeological properties of cave environments along with the effects of surface processes on cave morphology in the course of 50 years (Bekdemir, Sever, Uzun, & Elmacı, 2004; Kopar, 2009; Semenderoğlu, 2013; Semenderoğlu & Aytaç, 2013; Uzun, Zeybek, Yılmaz, & Bahadır, 2015). The impact of surface processes over cave structure was represented by an exemplary study in Central Taurus (Mediterranean Turkey). The observations of the study approached unroofing mechanism that transforms cave morphology in the context of exposure of vertical caves by cliffs, which is a rarely reported phenomenon in high mountains, as a result of the destruction of karst by glacial erosion in Aladaglar Massive (A. Klimchouk, Bayari, Nazik, & Törk, 2006). Important observations on regional hydrogeology and geomorphology as well as the evolution of Aladaglar Karst were reported, and the second deepest cave of Turkey in the date (Kuzgun Cave: 2080 m) was found within this study (Klimchouk, Nazik, Bayari, Törk, & Kasjan, n.d.).

Obruks of the Konya Closed Basin (KCB) (Central Anatolia), which are the particular type of dolines named after these examples in the international terminology, were studied to understand the structure and formation mechanisms (Bayarı, Pekkan, & Özyurt, 2009; Günay, Çörekçioğlu, Eroskay, & Övül, 2010). When the data gathered from hydrogeochemical analyses (depending on dissolution capacity of deep-seated waters) was interpreted considering the geology and hydrology of the area, it was concluded that the obruks found in the basin were formed by hypogenic mechanism (Bayarı et al., 2009; Özyurt & Bayarı, 2014). Moreover, the spatiotemporal distribution and the susceptibility of obruks within this plateau were evaluated by utilizing geographical information systems and geophysical methods (Özdemir, 2015, 2016). Collapse resistances and collapse conditions were focused on in the study conducted on obruks found in Sivas (Karacan & Yılmaz, 1997). Studies on the relation between caves and their connected aquifers are not limited to obruks only. The study that sheds light on hydrologic relation between İnsuyu Cave (Burdur, Turkey) and Çine Ovası Aquifer examined the change on the underground water table occurred due to the imbalance between recharge and discharge rates of the aquifer caused by the excessive pumping required to cover for increasing agricultural needs appeared because of the climate change. The effect of the semiarid climate effective in the region since 1986 on the underground water table is calculated through data gathered from VSR and direct measurements from wells, and the results could clearly be observed from the data gathered from these studies which are also important for shedding light on the mechanisms forming the unique pattern of the cave. In the cave that shows a complex structure as it is a combination of a network maze and big collapse galleries, these two morphologies are observed in two sections that cross from one to the other suddenly at a point close to the center of the cave (Taşdelen, 2018). It is a very rare condition for these two structures to be seen together. Detailed information gathered from the research focused on geomorphological structure of the some other caves mentioned in the context of this book and the related bibliography is given in Chap. 5.

Compared to terrestrial caves, geology and hydrology studies conducted in marine caves are limited in Turkey (Bayarı & Kurttaş, 2002; Bayarı, Özyurt, Hamarat, Bastanlar, & Varinlioglu, 2007; Coşkun, 1978; Elhatip & Günay, 1998; Elhatip, 2003; Hamarat, Ülkenli, Türe, & Bayarı, 1998). The most recent research was conducted by Bayarı and his team at the southwestern Mediterranean coast of Turkey (Bayarı et al., 2011). In this study, Patara–Kekova part of the coastal Western Taurus Region was examined in terms of freshwater discharges and waterfront-underwater caves. As an interdisciplinary approach, data related to geological, morphological, tectonic, and topographic features in addition to satellite imagery was used, reconnaissance diving to 0–30 m of depth was completed, in situ physical–chemical measurements were performed, chemical and isotopic compositions were determined through analysis in this study. As a result of the preliminary studies, it was determined that the examined area contained numerous freshwater discharges with the freshwater contribution up to 60% and 6 important anchialine underwater caves. Subsequently, Mivini and Altuğ Cave were profoundly studied to elucidate cave forming mechanisms and hydrodynamics of groundwater water discharge (Öztan et al., n.d., Özyurt, 2008). The study conducted in Gilindire Cave (İçel, Turkey), which is an important coastal cave and a tourist attraction, is also important for shedding light on the geomorphological evolution of the area (Nazik et al., 2001). The geomorphological pattern of the Gilindire Cave, which is situated in Cambrian aged dolomite and limestone lithology, reflects the forces that shaped the topography of the area on the south of the Mid-Taurus Karst Belt. The entrance of the cave found on the top part of a steep sea cliff is 46 m high from sea level, and the cave lies in three different sections developed in different periods with a total length of 555 m. These sections are comprised of the "entrance hall," which is composed of narrow galleries and cascaded levels situated on the NNW–SSE direction fault line, the "dripstone hall," which contains numerous speleothems composed of a kind of mineral sediment, and the "lake hall," the youngest section with dimensions of 140 m × 18–30 m. In this state, the general elongation of the cave is observed to have been affected by the structure of the unit it is found within, which is fractured by faults and shows tectonic dissections, the development of fracture systems, the layer directions and descents of the rocks, the climate, and the Pleistocene sea-level oscillations (Özşahin & Kaymaz, 2014).

In addition to provide critical information about the lithology and stratigraphy of the units, the tectonic evolution, hydrology and aquifer structures of the areas in which they situated, caves are raised the interests with high tourism potential. Caves, encountered in many cultures as shelters, sanctuaries, or burial places since the prehistoric period, also attract tourists' attention as historical remarks, geoheritage or thermal baths since the era of King Tiglath Pileser in 1100 BC (Cigna & Forti, 2013). By hundreds of examples around the World, touristic caves should meet a suit of condition including logistic convenience and structural stability. In this sense, studies are conducted to evaluate the stability of caves in Turkey as well. For example, Gilindire Cave was evaluated as a unique geoheritage with refers to its tourism potential with numerous cave features decorating the passageways (Özşahin & Kaymaz, 2014). In the Gökgöl Cave (Zonguldak, Turkey) and the Yazkonağı Cave (Ordu, Turkey) located in Black Sea Region, the lithology and tectonic structure with the fault-fracture system of the unit in which the cave developed were investigated to asses the stability of zones (Ersoy, Kirmaci, & Firat Ersoy, 2006; Geniş & Çolak, 2015). The Gökgöl Cave, in spite of receiving over 300,000 visitors since 2001, was reported with certain stability problems while the units cave developed within were in good quality, as a result of the examinations of geomechanical and physical properties of the cave and the numerical analysis (Geniş & Çolak, 2015). The radon level was also measured in the Gökgöl Cave in context of public health (Aytekin et al., 2006), due to being likely to be detected with high quantities in geologic formations such as caves, accumulate in closed environments and may cause lung cancer if it reaches critical values. For this reason, as in other countries (Solomon, Langroo, Lyons, & James, 1997; Szerbin, 1996), radon measurements in touristic caves are applied in Turkey. Radon measurements were also conducted in Karaca Cave in Gümüşhane, Çal Cave in Trabzon (Çevik et al., 2010), Tınaztepe Cave in Antalya (Çömlek, 2010) and Bulak Cave in Karabük (Haner, Yılmaz, Kürkçüoğlu, & Karadem, 2010). In another study conducted in the Bulak Cave, which is a 6 km long cave rich with speleothems and has a high potential for tourism (Y. Özdemir, 2005), the quality of air was also evaluated (Çetin, Sevik, & Saat, 2017).

3.3 The Research with Perspective of Speleothem Geochemistry

Cave sediments known as speleothems may be the top most attractive features of caves for the visitors. Speleothems in karstic caves are usually formed by the precipitation of calcium carbonates ($CaCO_3$). Meteoric water with dissolved CO_2 gas of the atmosphere is received by a basin as precipitation. While percolating through the soil zone, the waters absorb more CO_2 that is produced by the degradation of organic matters and a weak acid called carbonic acid (H_2CO_3) is formed. Carbonic acid causes dissolution of calcium carbonate ($CaCO_3$) rocks. The acidic surface waters become saturated with respect to calcium and enriched in carbonate (CO_{3-}) while leaking through the fissures of bedrock ($CaCO_3$). When seepage water (H_2O–CO_2–$CaCO_3$) saturated with calcite reaches the cold cave interior with less CO_2 partial pressure, CO_2 within dripping water releases into the cave atmosphere through degassing. At the same time, the water becomes supersaturated in bicarbonate (HCO_{3-}) and carbonate (Ca^{2+}) and calcium carbonate (CaCO3) is deposited by trapping the other free ions within the water that creates numerous subtypes in various colors and shapes (Ford & Williams, 2007) (Fig. 3.1). Speleothems may contain or be comprised of minerals including carbonates, silicates, nitrates, oxides and a variety of the other iron, magnesium, manganese and sulfur compounds (Hill & Forti, 1995). Besides, speleothem formation is also possible with ascending underground water by a similar mechanism in case of stability in high water level for a long time (e.g. anhydrite deposits, Dogtooth spar) or rapid evaporation (Fig. 3.1). Furthermore, the recent studies that reported microorganisms altering the rock geochemistry within cave environments brought a new aspect to the speleothem forming mechanism. The absence of solar radiation and primary production by photosynthesis within caves leads to alternative methods for organic carbon assimilation by the metabolic reactions referred to as chemosynthesis. Over years, the mounting evidence highlights the significance of chemosynthetic metabolism in bedrock dissolution/mineral precipitation processes within the context of microbe–mineral interactions (Barton & Northup, 2007; Carmichael, Carmichael, Santelli, Strom, & Bräuer, 2013; Engel et al., 2010; Legatzki et al., 2012). Mineralization mechanisms of speleothems are investigated by examination of speleothem morphology through visualization methods (e.g. SEM) as well as by detection of elements or markers that point out to factors, which may have a role in mineral precipitation, through geochemical and microbiological analyses (e.g. cultivation, isotopic measurement). Geochemical examination of a water sample gathered from the outermost layer of an active stalactite that continues its formation process, as in the study conducted in Küpeli Cave (Erdemli, Mersin, Turkey), represents an example for these methods.

Due to its formation mechanism, speleothem chemistry and structure is affected by (1) the source of the meteoric water involving the sediment precipitation as well as the topographic features and climate systems that the water encounters through its route until the cave area, (2) chemical properties and biological activities of the soil zone the water infiltrated through, (3) structural properties, lithology and geochemical characteristics of the bedrock (4) microclimatic properties of the cave ecosystem and (5) ongoing microbial activity in cave habitats (Fig. 3.1). Each of these elements leaves a chemical or physical fingerprint unique to itself. Being mostly isolated from the impact of short-term surface processes, speleothems with fingerprints like these may provide long-term data on surface and in-cave environmental conditions with good precision. Through the examination of cave sediments, studies that shed light on the past of micro and macro scale atmospheric events, climate trends, earthquakes and volcanic events in the nearby areas increase in number around the globe, while studies in this context are also conducted in Turkey (Baykara, 2014; Emery-Barbier & Thiébault, 2005; Göktürk et al., 2011; Rowe et al., 2012, 2020; Ünal-İmer, Shulmeister, Zhao, Uysal, & Feng, 2016).

Paleoclimate studies conducted with speleothems are highly popular globally, due to their quality of being high definition and well-protected long-term proxies (Baldini, Mcdermott, Hoffmann, Richards, & Clipson, 2008; Jones, Roberts, Leng, & Türkeş, 2006; Yang, Johnson, Griffiths, & Yoshimura, 2016). Many studies have been made on this subject also in Turkey, which has a complex climate regime developed under both global atmospheric teleconnections and regional topography. Turkey's climate regime, determined by air masses encountering atmospheric patterns. Turkey's climate, in general, is under the influence of the polar air mass during winter and the tropical air mass during the summer. Located in the Eastern Mediterranean Region, Turkey is affected by three major patterns of atmospheric teleconnections due to its geographical location. The first one is the North Atlantic Oscillation, which has an impact on winter. The North Atlantic Oscillation (NAS) is a large-scale meridional atmospheric circulation occurring under the influence of the climate system created by Icelandic Subpolar Low-Pressure Centre and the Azores Islands Subtropical High-Pressure Centre. The North Atlantic Oscillation, which affects the strength of the Western Atlantic Winds coming from the Mediterranean, increases the intensity of the Western winds in the negative phase and causes more humid and hot conditions in the Eastern Mediterranean (Göktürk, 2011). The North Sea-Caspian Pattern (NSCP), also taking effect in the winter period and generating from two constant pressure areas one of which is located in the North Sea and the other in the North of Caspian, is the second active climate component in Turkey. While this atmospheric connection pattern is in a positive phase—with the effect of Northern winds passing through Turkey—cold and dry weather conditions are observed (Göktürk, 2005). The last one is the influence of the monsoon (HM) originating in India. The summer climate regime in Turkey will be drier with the effect of India originating monsoon, which is an important part of the summer atmospheric circulation in the Northern Hemisphere.

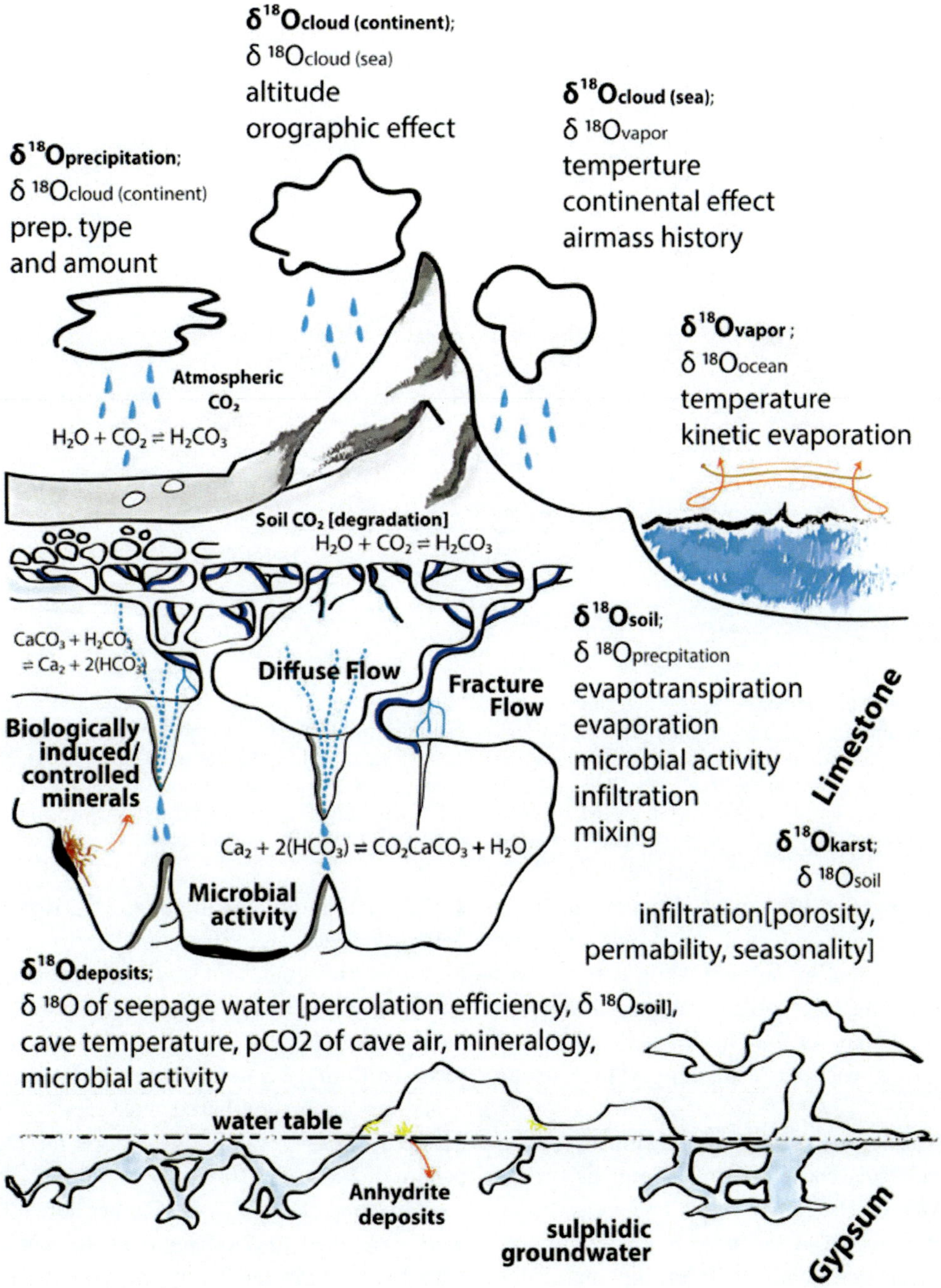

Fig. 3.1 Conceptual diagram of the atmosphere–surface–subsurface interactions and impacts within the perspective of cave environments. As an example of how landforms and landscape affect drip water and cave deposit geochemistry, the factors in the isotope fractionation of oxygen through the path from source to cave are described, herein. The mechanism of speleothem formation by meteoric water is given by the steps: (1) CO_2 adsorption from atmosphere and soil, (2) dissolution of limestone and enrichment of seepage water with Ca^{2+}, (3) cooling of seepage water, CO_2 degassing, and precipitation of $CaCO_3$ due to being water supersaturated. Two ways of water percolation through the rock are given as diffuse flow (generally in case of high permeability, more resistance to dissolution) and fracture flow. The other mechanisms of deposition are represented as biologically induced mineralization by microbial activity and mineral formation by acidic groundwater activity. Anhydrite deposits ($CaSO_4$) (illustrated with yellow sign, herein) are common features of hypogenic caves

Other than the aforementioned atmospheric teleconnection patterns and the motion of air masses, large mountain ranges and seas surrounding the three sides of the country affecting the climate conditions at an important scale, produced a complex climate regime that differentiates in each region (Fig. 3.2). The impact of the local elements on the active climate regime within the regions given; the north of the North Anatolian Mountain Belt, the southern part of the Taurus Mountain Belt and Central Anatolia, can be observed in speleothem records (Göktürk, 2011). For example, The North Anatolian Mountain belt forms as a natural barrier between the coastal Black Sea Region and the Central

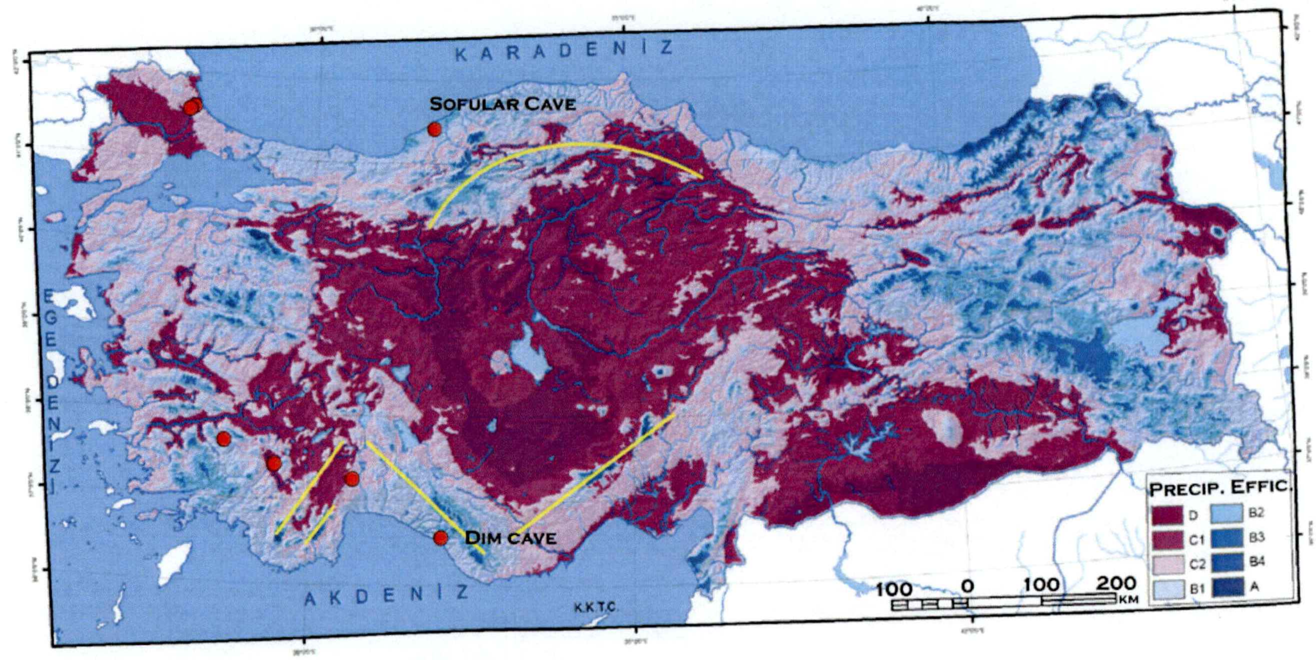

Fig. 3.2 Map of Turkey with the locations of the caves that investigated during the given paleoclimate researches: Göktürk (2011); Baykara (2014) and Ünal-İmer (2016). The relief basemap with the locations marked with red dots was prepared by E. Tok with QGIS v.3.12.0 software set the reference coordinate system to wgs84/UTM zone 36. The illustration that represents the precipitation efficiency zones of Turkey was modified from Yılmaz and Çiçek (2016). The colors are indicating the different rates of precipitation efficiency that are defined by Thornthwaite method. A: Perhumid; B1, B2, B3, B4: humid; C1: dry subhumid; C2: moist subhumid; D: semiarid. The yellow lines are indicating the mountain ranges; the Black Sea Mountains in the north and the Taurus Mountains in the south

Anatolia during the invasion of warm and humid maritime air that was initially a dry and cool continental air mass coming from the North that humidified while passing over the Black Sea. The moisture content releases as intense rainfall on the coastal slope of the North Anatolian Mountains while the air mass reaches the interior part of Anatolia as dry and warm.

This has led to discrepancies between the climate records from Turkey's Black Sea coasts and other regions on the impact of North Atlantic Oscillation over these areas. In the study conducted in Sofular Cave situated in this region, the $\delta^{18}O$ isotope composition attained from the stalagmite samples revealed the distinct nautical effect of the Black Sea on the local climate regime. It was seen that the $\delta^{18}O$ isotope composition attained from the stalagmite samples gathered from the cave was consistent with the data gathered from previous studies conducted in the Black Sea Region but, it didn't show any meaningful correlation with the general isotope profile that represents the North Atlantic Oscillation. As a result of comparison between similar studies conducted on cave sediments found in the East Mediterranean, records parallel to climatic processes effective on large areas were observed and it was remarked that the records were regional in fast climatic events due to the Black Sea's nautical effect. Furthermore, while the stable isotope $\delta^{13}C$ analysis and the radiochronologic dating methods point to the photosynthetic activity and the flora with dense $\delta^{13}C$ isotope values, which belongs to Early Holocene Period and was gathered from the examined stalagmites, in studies conducted in other regions, it was remarked that this distinct increase in flora started at a later period (Göktürk et al., 2011). Again, speleothems gathered from the Sofular Cave have been examined through synchrotron-radiation-based trace element analysis and the Minoan volcanic eruption's effect on the region, which had occurred between 1600 and 1650 BC, was researched (Badertscher et al., 2014). Furthermore, these stalagmites were examined to shed light on the timing and effect of Dansgaard–Oeschger events in Turkey (Fleitmann et al., 2009). Through the increasing number of speleothem researches, and thus the understanding of regional parameters that cause differences over proxies, these sediments that bear the qualities of well preserved, long-term climate archives would play a huge role in revealing Anatolia's detailed climate history.

Another component of orographic and continental effect that exerts a great importance on regional climate differentiation in Turkey is the Taurus Mountains (Fig. 3.2). Strong convectional rains are seen frequently in the region that generally has a mild and humid climate. The low value of the $\delta^{13}C$ isotope rate is accepted as the indicator in climate records for mild and humid climate regime that causes kinetic fractionation in countenance for light isotope through the enrichment of surface fauna and the increase in photosynthetic activity that comes with it. The results of the study conducted in Kocain Cave (Antalya, Turkey) point out to higher $\delta^{13}C$ isotope amount than expected in this region where C3-type vegetation and Mediterranean climate is dominant. The reason for this is shown to be low vegetation cover and ineffective infiltration, in the study conducted in Sofular Cave and the study conducted as a part of Göktürk's Ph.D. thesis presented in 2011 (Göktürk, 2011). The atmospheric teleconnection patterns, which were active during Holocene and the effect of local variables in the climatic conditions of the Eastern Mediterranean, were investigated. Efforts were made to interpret the regional effects of paleoclimatic processes in Turkey from speleothem records and to support the examinations being conducted in Uzuntarla and Yenesu Cave found in Thrace, Ovacık Cave found in Black Sea Region, Sofular Cave and Kocain Cave (Göktürk, 2011). In the research presented by Wickens as a Ph.D. thesis study in 2013, the reason for the geochronologic deviation seen on the stalagmite sample gathered from the study conducted in Dim Cave (Antalya, Turkey) found South of the Taurus Mountains is shown to be the local high rainfall that led to flushing events and recrystallization (Wickens, 2013). Within this study, the changes in the climate trend during the Permian Period (5 AD) were observed clearly. Another study conducted in Dim Cave presented the climate changes in the region of the last 80,000 years with high definition oxygen isotope data (Ünal-İmer et al., 2015, 2016). This study that traces the signs of the events included by eastern Mediterranean's climate history in this region, revealed that the fluctuations in $\delta^{18}O$ of speleothem record represent a correlation with that of the oceanic moisture source driven by latitudinal shifts in the westerlies. Within the Ph.D. thesis of Baykara in 2014, stalagmites gathered from Sırtlanini, Keloğlan and Dim Cave were aged by the U-Th method and formation processes of those were examined. In this study, the Permian Period and Holocene paleoclimate conditions were tried to be shed light on by the proxies of the stable isotopes (^{18}O and ^{13}C) (Baykara, 2014).

Earthquakes that made an effect in the region throughout history, are another ground system process that stalagmites show long-term record qualities over. Earthquakes may cause deviations on stalagmites growth axis by changing the dripping point of the dripping water that causes calcite accumulation during the stalagmite formation or by directly causing cracks and fractures on stalagmites. Chronologically compatible and angularly similar deviations detected on stalagmite samples gathered from the same cave or areas close by can point out to a past earthquake that affected the area when interpreted together with data like lamination thickness and color. As it has been hosting communities that shaped civilization throughout archaeological periods, shedding light on tectonic events occurred in Anatolia, bears great importance for the history of lands extending from Europe to Asia. Traces of three big earthquakes that occurred in the Mediterranean between 110 and 360 AD were detected on a stalagmite sample gathered from Kepez Cave (Mersin) (Akgöz & Eren, 2015). Bayarı and Özyurt (2005) point out that those stalagmite growth axis changes that depend on the changes of the primary stalagmite position with time are caused by tectonic raise or subsidence of the region. In this sense, speleothems provide opportunities to shed light on the history of tectonic processes through systematic studies conducted on regions related to specific tectonic systems.

One of the most important analyses applied in these aforementioned researches is dating. Chemical and biologic data gathered from various levels of the sediments are placed in an order and a timeframe so that the data pointing out to natural processes occurred on earth's crust and the atmosphere can be interpreted in historical context. In this sense, studies examining various methods used in the dating of cave sediments were also conducted. The study Engin and his team (2010) conducted in Keloğlan Cave (Acıpayam, Denizli, the southwestern part of Turkey) researched the usability of the ESR dating method.

3.4 The Research with Perspective of Cave Microbiology

Even though usually the $\delta^{18}O$ and $\delta^{13}C$ isotopes and trace elements are examined in paleoclimate studies, microorganisms protected in cave sediments also function as biomarkers for climatic conditions (Epure, 2015). Microclimatic conditions (which may be affected by long-term atmospheric conditions), along with composition and concentration of elements/nutrients available for microbes in cave environments, supports some particular niches and leads to a differentiation in the abundance and composition of the microbial communities (Schabereiter-Gurtner et al., 2004; Engel and Northup, 2008). Studies exhibits that cave conditions not suitable for photosynthetic life support surprisingly splendid chemosynthetic microflora (Tomczyk-Zak & Zielenkiewicz, 2016). Mineralization, during the formation of speleothems like stalactites, stalagmites and other cave sediments, can occur inorganically or biologically induced/controlled with the presence of the certain chemosynthetic microorganisms. Even if the early studies attributed the microbial life in cave environments only to heterotrophic consumption of organic carbon transported into caves from surface, surrounding rock or overlaying soil zone (Engel, Porter, Stern, Quinlan, & Bennett, 2004), the recent studies prove that the ecological function of microbial communities introduce diversity far more than that (Northup & Lavoie, 2001; Barton & Northup, 2007). Bacterial cave microflora is commonly composed of sulfur-oxidizing, iron-oxidizing, hydrogen-oxidizing taxa and taxa that play a part in the nitrite–nitrate cycle (Engel, 2015). Well-known with high adaptation capability, the classes Alphaproteobacteria, Gammaproteobacteria, Betaproteobacteria, Deltaproteobacteria and the groups Acidobacteria, Nitrospirae, Actinobacteria are the most encountered taxa in cave environments (Tomczyk-Zak & Zielenkiewicz, 2016). The major dominant group Actinobacteria is attributed to acicular aragonite/calcite crystals which develop on mats (Barton et al., 2004; Canaveras et al., 1999; Canaveras et al., 2001). It is known that some species belonging to Alphaproteobacteria class play a role in iron oxidation (*Hyphomicrobium* spp. *Pedomicrobium* spp. etc.), in nitrogen fixation (*Mesorhizobium* spp.), and in methane cycle (*Methylobacteria* sp.) (Barton et al., 2004; Carmichael et al., 2013). Some members of this class (Spingomonas) exerts an important role in biomineralization processes through the production of Extracellular Polymeric Substance (EPS) by metabolizing various aromatic compounds in caves (Canaveras et al., 1999; Canaveras et al., 2001; Northup & Lavoie, 2001). Betaproteobacteria class is represented by species that oxidize sulfur (*Thiobacillus* spp., *Thiobacter* spp.), nitrogen (*Nitrosomonas europaea*, *Nitrosomonas oligotropha*) and nitrate (Nitrotoga arctica) as well as methylotrophic (*Methylotenera* spp., *Methylophilus* spp., *Methylovorus* spp.) and denitrifying (*Denitratisoma* spp.) (Tomczyk-Zak & Zielenkiewicz, 2016). Besides the members of Gammaproteobacteria that oxidize sulfur are frequently reported in cave environments rich in oxygen and sulfur, some of the taxa (Pseudomonas spp.) are linked to ferromanganese deposits (Carmichael et al., 2013). The level of $O_{2(g)}$ in cave air exercises a critical role over niche differentiation of the species with sulfur-oxidizing metabolism of the class Epsilonproteobacteria, where thrive in oxygen limited environments (Macalady et al., 2008). Thus, it may be assisted that there is a two-way microbe–mineral relation in cave environments.

The distribution and abundance of cave microorganisms in Turkey represents a similar composition with the karstic caves around the World (Atakav, 2017; Güleçal-Pektaş & Temel, 2017; Tok, 2017). One of the first research focused on cave microflora was investigated 19 karstic caves located in different regions of Turkey (Yamaç, Işik, & Şahin, 2011). In the extent of this study, the distribution of streptomyces genus in Actinobacteria phylum was examined in these caves. Some of these streptomyces isolates were selected to be evaluated in terms of their antimicrobial activities (Yücel & Yamaç, 2010). Tillo Busto (Spain) and Cervi Cave (Italy) are examples of caves with rich Actinobacteria population (Groth, Vettermann, Schuetze, Schumann, & Saiz-Jimenez, 1999; Laiz et al., 2000). In a study conducted to examine the microbial diversity in İnsuyu Cave (Burdur) (Figs. 3.3 and 3.4), the most common class was observed to be the Alphaproteobacteria (% 89.23 of the total bacteria) (Tok, 2017). This class was followed by Actinobacteria (%3.94), Bacilli (% 2.92), Gammaproteobacteria (%1.69) and Betaproteobacteria (%0.74). The detected fourteen genus were identified as Methylobacterium (%88.83), Propionibacterium (%3.58), Dolosigranulum (%1.69), streptococcus (%1.10) and Pseudomonas (%1.13). Furthermore, the morphologies of the samples and dispersion of the population in this study were compared to the results gathered from other studies conducted in foreign countries. In another study examining the cave bacteria flora in Gilindire Cave (Mersin), the distribution was identified as Actinobacteria (39%), Proteobacteria (%33), Firmicutes (17%) and Bacteroidetes (5.5%) in addition to uncultured organisms (5.5%) (Atakav, 2017).

Fig. 3.3 Photograph from İnsuyu Cave, representing a sampling location of the research by Tok (2017), taken by A. Yamaç

3.5 The Research with Perspective of Cave Fauna

Besides the microflora, macro-organisms like bats, fish, insects and spiders in cave environments were also examined by biological studies in Turkey. First known biospeleology research is widely attributed to Macarlı Miralay Dr. Abdullah

Fig. 3.4 Sediment photographs from İnsuyu Cave. **a** Black, velvet-textured deposit. **b** Acicular crystal deposit. **c** Preparation to laboratory analysis (Tok, 2017)

Bey with refer to his study in Yarımburgaz Cave (İstanbul, Turkey) in 1865 (Kunt, Yağmur, Durmuş, & Anlaş, 2010). On the other hand, the first study conducted with a systematic approach was conducted by the Spanish naturalist and insectologist Ignacio Bolivar on samples gathered by his colleague M. Martinez, who visit to Anatolia for his own studies. Among almost 100 different species of arthropods which were collected from the area that extends from İskenderun to Maraş and from the Taurus Mountains to the Binboğa Mountains, the species gathered from Yenicekale (Kahramanmaraş, Turkey) and Akbaş Caves (İskenderun, Turkey) are the very first macro-organisms sampled from

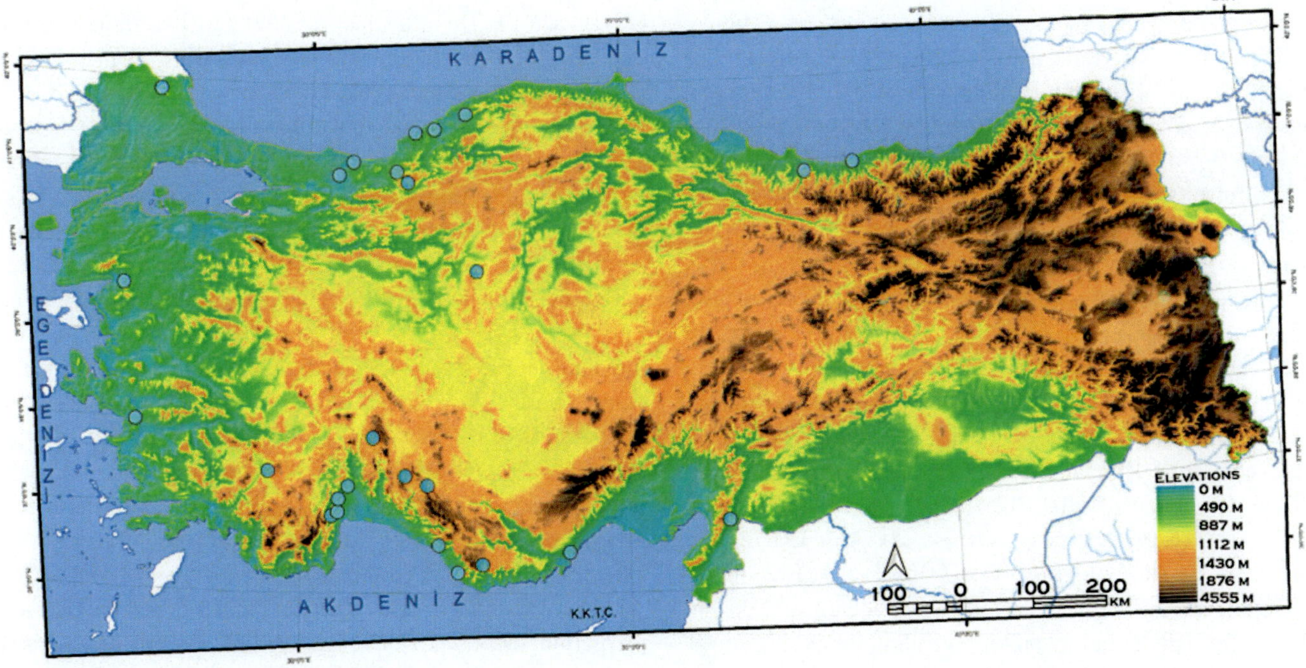

Fig. 3.5 Map of Turkey with the locations of the caves that investigated during the given arthropod researches: Di Russo, Rampini & Landeck (2007); Külköylüoğlu, Yavuzatmaca, Karacaoğlu & Telli (2014); Kunt, Özkütük, Elverici, Marusik, & Karakaş (2016); Ribera, Elverici, Kunt & Özkütük (2014); Taylan, Di Russo, Rampini, & Cobolli (2011); Taylan, Yılmazer & Şirin (2020). The relief basemap with the locations marked with blue dots was prepared by E. Tok with QGIS v.3.12.0 software set the reference coordinate system to wgs84/UTM zone 36. The colors are representing the elevations from sea level

cave environments in Anatolia. Primarily focused on cave crickets and arachnoids, there have been many arthropod studies conducted in various parts of Turkey since then (Taylan, Di Russo, Rampini, & Cobolli, 2011) (Fig. 3.5).

In one of these researches, the individuals identified as Discoptila beroni (Insecta: Ensifera: Gryllidae) were observed larger in body sizes within the Yalan Dünya Cave (Antalya, the southern of Turkey) than the ones detected in the nearby area (Köhler, Renker, & Luis, 2004). Among the two new species identified in another study conducted in Artvin (the Black Sea Region of northeast Turkey) in 2007, *Dolichopoda noctivaga* sp. n. was separated from the other individuals encountered in the area by the shape of its epiphallus, while the smaller body size compared to the ones nearby was the differentiating feature for *Troglophilus tatyanae* sp. n. (Di Russo, Rampini, & Landeck, 2007). A study on cave microclimate and distribution of cave cricket species was conducted by Taylan and his team (2020). In this study, three caves were selected from three different regions of Turkey; Geyikbayırı Cave (Antalya, the southern of Turkey), Sipahiler Cave (Bartın, the Black Sea Region of Turkey), Tuluntaş Cave (Ankara, central Turkey) were compared in terms of temperature, moisture and the climate regimes of the regions in which the caves are situated. The relationship between surface climate and microclimate was assessed and the impact of these environmental conditions on the distribution of cave crickets was interpreted. The annual mean temperature is the highest in Geyik Bayırı Cave, located in the Mediterranean region that is characterized by hot, moderately dry summers and mild to cool, wet winters; while it is the lowest at Sipahiler Cave located in the Black Sea Region with warm, wet summers and cool to cold, wet winters. Also, the relative humidity of cave zones in Geyikbayırı Cave is higher than that of its local climate regime. The region Tuluntaş Cave is located in, is described with continental climate regime with sharply contrasting seasons in temperature. The research assessed the cricket populations of these caves whether a correlation exists with these conditions while the other parameters are also taken into consideration such as vegetation cover and elevation (Taylan, Yılmazer, & Şirin, 2020). Aside from locusts, arachnoids (Kunt, Özkütük, Elverici, Marusik, & Karakaş, 2016; Ribera, Elverici, Kunt, & Özkütük, 2014) millipede (Antić, Çetin, Turantepe, & Gürbüz, 2016; Enghoff, 2006), amphipoda (Özbek, 2007, 2012a, 2012b; Özbek & Oktar Guloglu, 2005; Özbek, Yurga, & Külköylüoğlu, 2013; Özkan, 2009), fishes (Özdemir & Erkakan, 2014), ostracoda and planktons (Külköylüoğlu, Yavuzatmaca, Karacaoğlu, & Telli, 2014) were examined in various studies.

Bats composing of a large part of cave dwellers are studied all over the country (Fig. 3.6), with the records of *Barbastella barbastellus*, *Eptesicus serotinus*, *Hypsugo savii*, *M. bechsteinii*, *M. brandtii*, *M. daubentonii*, *M. nattereri*, *Miniopterus schreibersii*, Myotis blythii, *Myotis capaccinii*, *Myotis emarginatus*, *Myotis myotis*, *Myotis mystacinus*, *Nyctalus leisleri*, *Pipistrellus kuhlii*, *Pipistrellus pipistrellus*, *Plecotus austriacus*, *Plecotus auritus*, *R. meheyli*, *Rhinolophus euryale*, *Rhinolophus ferrumequinum*, *Rhinolophus hipposideros*, *Tadarida teniotis* (Albayrak, 2003 gathered from the studies: Albayrak, 1990; Aşan, Baydemir & Albayrak, 2006; Aşan & Albayrak, 2011; Benda & Horacek, 1998; Helversen, 1989; Steiner & Gaisler, 1994). One of the most extensive studies on cave-dwelling bats in Turkey was conducted by Furman and Özgül (2002, 2004) in Eastern Thrace in 1999–2000 and in Western Thrace in 2001. From eight underground sites in Eastern Thrace, 17,000 bat records representing eight species were covered, while there were approximately 76,000 bat records, representing 13 species from 32 underground sites covered from the western extension. Koyunbaba and Dupnisa Caves were reported to be the most important caves for bat dwelling, hibernaculum and nursery. The authors also raised their concern about the touristic activity that may harm the bat population in the Dupnisa Cave System (Kırklareli, the Thrace region of northwest Turkey).

Within this context, another study that investigated the seasonal population dynamics following the opening of Dupnisa Cave for tourism to understand the impact of visitors and show cave arrangements by a total of 53 surveys, reported no negative consequences observed (Paksuz & Özkan, 2012). The seasonal variations of bat fauna and the effect of microclimate on bat population were also studied in Dupnisa Cave System (Paksuz, Özkan, & Postawa, 2008). The study revealed that the different parts of the Dupnisa Cave System (Sulu Cave, Kuru Cave, Kız Cave) serve the purposes of dwelling, hibernaculum and nursery differently. While the particular temperature of each part was found to be in correlation with the species composition, no correlation with humidity was observed. The caves with bat populations in Northwest of Central Anatolia were investigated to survey the distribution of bat species and the characteristics of these caves (Barlas & Yamaç, 2016). In different studies conducted between 2016 and 2010 about bat diversity in Turkey, different protection unit approaches (evolutionarily important unit, protection management unit and population grouping analysis) were examined for the protection of genetic diversity of *Rhinolophus ferrumequinum*, *Miniopterus schreibersii* and *Myotis capaccinii* and cave based protection strategies were suggested as a result of analyses applied for the first time for species living in Turkey (Bilgin, 2012). Giant Wings of the Underground Project is an exemplary work among the studies that aim for the detection of bat groups, preparation of cave protection plans for the

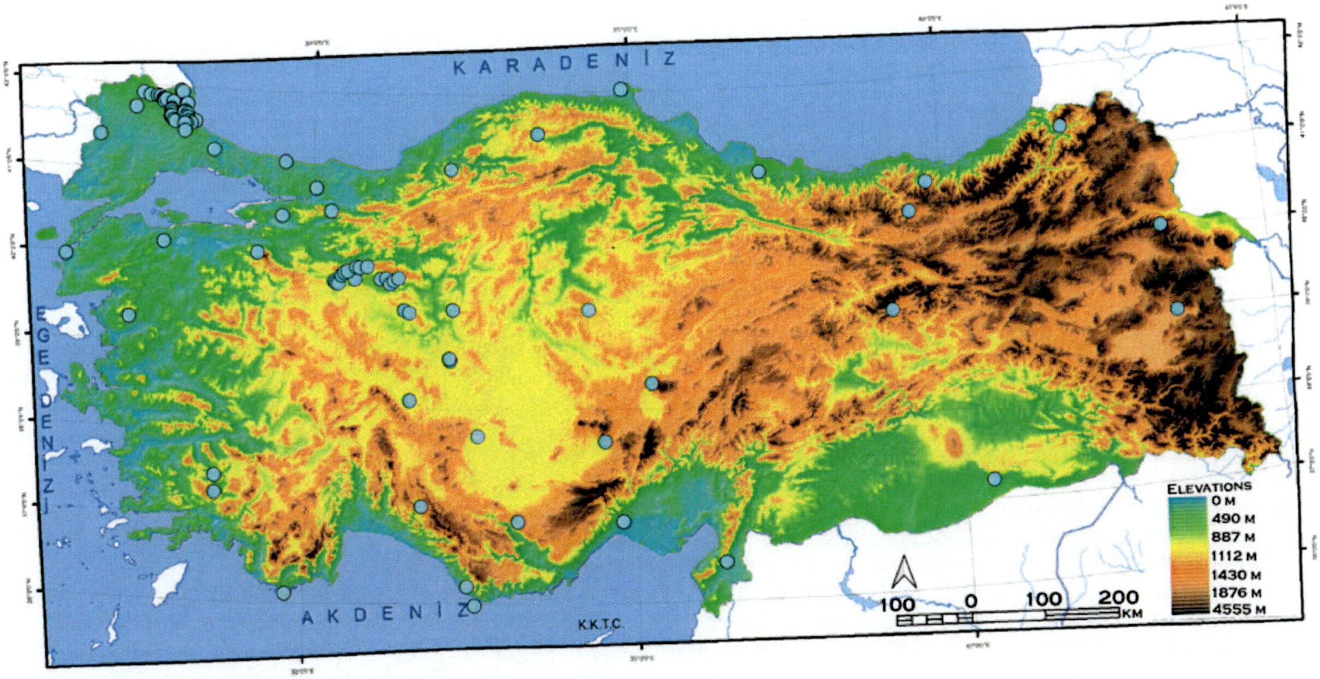

Fig. 3.6 Map of Turkey with the locations of the caves that investigated during the given cave-dwelling bat researches: Barlas and Yamaç (2016); Bilgin (2012); Furman and Özgül (2002). The relief basemap with the locations marked with blue dots was prepared by E. Tok with QGIS v.3.12.0 software set the reference coordinate system to wgs84/UTM zone 36. The colors are representing the elevations from sea level

protection of cave ecosystems and ensuring that protective legal precautions are taken. Studies are still being conducted to detect the population size and protection needs of the Egyptian fruit bat (*Rousettus aegyptiacus*), which can only be observed in Turkey in caves found in the East Mediterranean and the Mid-Mediterranean regions.

References

Akgöz, M., & Eren, M. (2015). Traces of earthquakes in the caves: Sakarlak Ponor and Kepez Cave, Mersin (S Turkey). *Journal of Cave and Karst Studies, 77*(1), 63–74.

Albayrak, İ. (2003). The bats of the eastern Black Sea region in Turkey (Mammalia: Chiroptera). *Turkish Journal of Zoology*.

Albayrak, İ. (1990). Doğu Anadolu yarasaları ve yayılışları (Mammalia; Chiroptera). *Turkish Journal of Zoology, 2*(14), 214–228.

Antić, D., Çetin, C. Ç., Turantepe, E., & Gürbüz, M. F. (2016). A new cave-dwelling species of the genus Mesoiulus Berlese, 1886 (Diplopoda, Julida, Julidae) from Turkey. *Zootaxa*. https://doi.org/10.11646/zootaxa.4196.2.8

Aşan Baydemir, N., & Albayrak, I. (2006). A study on the breeding biology of some bat species in Turkey (Mammalia: Chiroptera). *Turkish Journal of Zoology*.

Aşan, N., & Albayrak, I. (2011). Taxonomic status of Myotis myotis (Borkhausen, 1797) and Myotis blythii (Tomes, 1857) in Turkey (Mammalia: Chiroptera). *Turkish Journal of Zoology, 35*(3), 357–365.

Aşan Baydemir, N., & Albayrak, I. (2011). Taxonomic status of *Myotis myotis* (Borkhausen, 1797) and *Myotis blythii* (tomes, 1857) in Turkey (Mammalia: Chiroptera). *Turkish Journal of Zoology*. https://doi.org/10.3906/zoo-0803-4.

Aşan Baydemir, N., & Tüzmen, Ş. (2016). Cave ecosystems of Turkey and Northern Cyprus: A Hidden world for bats. *Journal of International Environmental Application and Science*.

Atakav, Y. (2017). Isolation, characterization and identification of bacteria live in extreme conditions/Ekstrem koşullarda yaşayan bakterilerin izolasyonu, karakterizasyonu ve identifikasyonu.

Aytekin, H., Baldık, R., Çelebi, N., Ataksor, B., Taşdelen, M., & Kopuz, G. (2006). Radon measurements in the caves of Zonguldak (Turkey). *Radiation Protection Dosimetry*. https://doi.org/10.1093/rpd/nci326.

Badertscher, S., Borsato, A., Frisia, S., Cheng, H., Edwards, R. L., Tüysüz, O., et al. (2014). Speleothems as sensitive recorders of volcanic eruptions—The Bronze Age Minoan eruption recorded in a stalagmite from Turkey. *Earth and Planetary Science Letters*. https://doi.org/10.1016/j.epsl.2014.01.041.

Badertscher, S., Fleitmann, D., Cheng, H., Edwards, R. L., Göktürk, O. M., Zumbühl, A., … Tüysüz, O. (2011). Pleistocene water intrusions from the Mediterranean and Caspian seas into the Black Sea. *Nature Geoscience*. https://doi.org/10.1038/ngeo1106

Baldini, J. U. L., Mcdermott, F., Hoffmann, D. L., Richards, D. A., & Clipson, N. (2008). Very high-frequency and seasonal cave atmosphere PCO2 variability: Implications for stalagmite growth and oxygenisotope-based paleoclimate records. *Earth and Planetary Science Letters, 272*, 118–129.

Barlas, E., & Yamaç, E. (2016). Cave ecosystems of Turkey and Northern Cyprus: A hidden world for bats. *Journal of International Environmental Application and Science*.

Barton, H. A., & Northup, D. E. (2007). Geomicrobiology in cave environments: past, current and future perspectives. *Journal of Cave and Karst Studies, 69*, 163–178.

Barton, H. A., Taylor, M. R., & Pace, N. R. (2004). Molecular phylogenetic analysis of a bacterial community in an oligotrophic cave environment. *Geomicrobiology Journal, 21*(1), 11–20.

Bayari, C. S., & Kurttaş, T. (2002). Coastal and submarine karstic discharges in the Gökova Bay, SW Turkey. *Quarterly Journal of Engineering Geology and Hydrogeology*. https://doi.org/10.1144/1470-9236/01034.

Bayarı, C. S., & Özyurt, N. N. (2005). Konya Kapalı Havzası yeraltı suyunda karbon-14 yaş dağılımı, II. Ulusal Hidrolojide İzotop Teknikleri Sempozyumu 26–30 Eylül 2005 Gümüldür-İzmir Bildiriler Kitabı, 147–168.

Bayarı, C. S., Özyurt, N. N., Hamarat, S., Bastanlar, Y., & Varinlioğlu, G. (2007). *Recovery of fresh water discharges along Turkish coast: Patara- Tekirova Pilot Project. TUBITAK Project Report no CAYDAG- 103Y025TUBITAK Project Report no CAYDAG-103Y025*. Ankara, Turkey.

Bayarı, C. S., Özyurt, N. N., Oztan, M., Bastanlar, Y., Varinlioglu, G., Koyuncu, H., … Hamarat, S. (2011). Submarine and coastal karstic groundwater discharges along the Southwestern Mediterranean coast of Turkey. *Hydrogeology Journal*. https://doi.org/10.1007/s10040-010-0677-y.

Bayarı, C. S., Pekkan, E., & Özyurt, N. N. (2009). Obruks, as giant collapse dolines caused by hypogenic karstification in central Anatolia, Turkey: Analysis of likely formation processes. *Hydrogeology Journal*. https://doi.org/10.1007/s10040-008-0351-9.

Baykara, M. O. (2014). Güneybatı Anadolu'da mağara çökellerinin incelenmesi ve paleoiklimsel değerlendirmeleri. Pamukkale Üniversitesi.

Bekdemir, Ü., Sever, R., Uzun, A., & Elmacı, S. (2004). Yıldızkaya Mağarası/Yıldızkaya Cave. *Doğu Coğrafya Dergisi*. https://doi.org/10.17295/dcd.56033.

Benda, P., & Horacek, I. (1998). Bats (Mammalia: Chiroptera) of the eastern Mediterranean. Part 1. Review of distribution and taxonomy of bats in Turkey. *Acta Societatis Zoologicae Bohemicae*.

Bilgin, I. R. (2012). The conservation genetics of three cave-dwelling bat species in southeastern Europe and Anatolia. *Turkish Journal of Zoology*. https://doi.org/10.3906/zoo-1012-74.

Canaveras, J. C., Hoyos, S. M., Sanchez-Moral, E., Sanz-Rubio, J., Bedoya, V., Soler, I., Groth, P., Schumann, L., Laiz, I., Gonzalez, C., & Jimenez, S. (1999). Microbial communities associated with hydromagnesite and needle-fiber aragonite deposits in a karstic cave (Altamira, Northern Spain), *Geomicrobiology Journal*, *16*(1), 9–25.

Canaveras, J. C., Sanchez-Moral, S., Soler, V., & Saiz-Jimenez, C. (2001). Microorganisms and microbially induced fabrics in cave walls, *Geomicrobiology Journal*, *18*(3), 223–240.

Carmichael, M. J., Carmichael, S. K., Santelli, C. M., Strom, A., & Bräuer, S. L. (2013). Mn(II)-oxidizing bacteria are Abundant and Environmentally Relevant Members of Ferromanganese Deposits in Caves of the Upper Tennessee River Basin, *Geomicrobiology Journal*, *30*(9), 779–800.

Çetin, M., Sevik, H., & Saat, A. (2017). Indoor air quality: The samples of Safranbolu Bulak Mencilis cave. *Fresenius Environmental Bulletin*.

Çevik, U., Kara, A., Çelik, N., Karabudak, S. M., & Celik, A. (2012). Radon survey and exposure assessment in Karaca and Çal Caves, Turkey. *Water Air and Soil Pollution*, *214*, 461–469. https://doi.org/10.1007/s11270-010-0437-6.

Cigna, A. A., & Forti, P. (2013). Caves: The most important geotouristic feature in the world. *Tourism and Karst Areas, 6*(1), 9–26.

Çömlek, Ü. (2010). *Tınaztepe Mağarası'nın radon seviyesinin araştırılması*. Selçuk University.

Coşkun, N. (1978). *Hydrogeological survey report of sea springs of Antalya-Kas-Kalkan region. DSI open file report*.

Di Russo, C., Rampini, M., & Landeck, I. (2007). The cave crickets of northeast Turkey and transCaucasian regions, with descriptions of two new species of the genera *Dolichopoda* and *Troglophilus* (Orthoptera, *Rhaphidophoridae*). *Journal of Orthoptera Research*. https://doi.org/10.1665/1082-6467(2007)16%5b67:tccont%5d2.0.co;2.

Elhatip, H., & Günay, G. (1998). Karst hydrogeology of the Kas-Kalkan springs along the Mediterranean coast of Turkey. *Environmental Geology*. https://doi.org/10.1007/s002540050330.

Elhatip, H. (2003). The use of hydrochemical techniques to estimate the discharge of Ovac1k submarine springs on the Mediterranean coast of Turkey. *Environmental Geology*. https://doi.org/10.1007/s00254-002-0668-y.

Emery-Barbier, A., & Thiébault, S. (2005). Preliminary conclusions on the Late Glacial vegetation in south-west Anatolia (Turkey): The complementary nature of palynological and anthracological approaches. *Journal of Archaeological Science*. https://doi.org/10.1016/j.jas.2005.03.006.

Engel, A. S. (2004). Bacterial diversity and ecosystem function of filamentous microbial mats from aphotic (cave) sulfidic springs dominated by chemolithoautotrophic "Epsilonproteobacteria". *FEMS Microbiology Ecology*, *51*(1), 31–53.

Engel, A. S. (2015). Microbial Life of Cave Systems, Life in extreme environments, Walter de Gruyter GmbH & Co KG, p. 350.

Engel, A. S., & Northup, D. E. (2008). Caves and karst as model systems for advancing the microbial sciences. *Frontiers in Karst Research*, 37–48.

Engel, A. S., Meisinger, D., & Porter, M. (2010). Linking phylogenetic and functional diversity to nutrient spiraling in microbial mats from Lower Kane Cave (USA). *The ISME journal*, *4*(1), 98–110.

Enghoff, H. (2006). The millipedes of Turkey (*Diplopoda*). *Natural History*.

Engin, B., Aydaş, C., Özkul, M., Zeyrek, C. T., Büyüm, M., & Gül, A. (2010). ESR dating of stalagmite from Keloğlan cave in the southwestern part of Turkey. *Radiation Effects and Defects in Solids*, *165*(12), 944–957.

Epure, L., Muntean, V., Constantin, S., & Moldovan, O. T. (2015). Ecophysiological groups of bacteria from cave sediments as potential indicators of paleoclimate, *Quaternary International*, *1*, 1–13.

Ersoy, H., Kırmacı, M. Z., & Ersoy, A. F. (2006). Geology and formation of Yazkonağı Cave (Ünye-Ordu). *Jeoloji Muhendisligi Dergisi*.

Fleitmann, D., Cheng, H., Badertscher, S., Edwards, R. L., Mudelsee, M., Göktürk, O. M., … Tuysuz, O. (2009). Timing and climatic impact of Greenland interstadials recorded instalagmites from northern Turkey. *Geophysical Research Letters*, *36*. https://doi.org/10.1029/2009GL040050.

Ford, D., & Williams, P. (2007). *Karst hydrogeology and geomorphology*, West Sussex: John Wiley & Sons Ltd.

Furman, A., & Özgül, A. (2002). Distribution of cave-dwelling bats and conservation status of underground habitats in the Istanbul area. *Ecological Research*. https://doi.org/10.1046/j.1440-1703.2002.00468.x.

Furman, A., & Özgül, A. (2004). The distribution of cave-dwelling bats and conservation status of underground habitats in Northwestern Turkey. *Biological Conservation*. https://doi.org/10.1016/j.biocon.2004.02.019.

Geniş, M., & Çolak, B. (2015). Stability assessment of the Gökgöl Karstic Cave (Zonguldak, Turkey) by analytical and numerical methods. *Rock Mechanics and Rock Engineering*. https://doi.org/10.1007/s00603-014-0700-z.

Göktürk, O. M., Fleitmann, D., Badertscher, S., Cheng, H., Edwards, R. L., Leuenberger, M., … Kramers, J. (2011). Climate on the southern Black Sea coast during the Holocene: Implications from the Sofular Cave record. *Quaternary Science Reviews*. https://doi.org/10.1016/j.quascirev.2011.05.007

Göktürk, O. M. (2005). *North Sea–Caspian Pattern and its influence On the hydrometeorological parameters over Turkey*. İstanbul Technical University.

Göktürk, O. M. (2011). *Climate in the Eastern Mediterranean through the Holocene inferred from Turkish stalagmites*. University of Bern.

Groth, I., Vettermann, R., Schuetze, B., Schumann, P., & Saiz-Jimenez, C. (1999). Actinomycetes in Karstic caves of northern Spain (Altamira and Tito Bustillo). *Journal of Microbiological Methods*. https://doi.org/10.1016/S0167-7012(99)00016-0.

Gulecal-Pektas, Y., & Temel, M. (2017). A Window to the Subsurface: Microbial diversity in hot springs of a sulfidic cave (Kaklik, Turkey). *Geomicrobiology Journal, 34*, 374–384.

Günay, G., Çörekçioğlu, Eroskay, S. O., & Övül, G. (2010). Konya Karapınar obruks (Sinkholes) of Turkey. *Environmental Earth Sciences*. https://doi.org/10.1007/978-3-642-12486-0_57.

Günay, G., Güner, N., & Törk, K. (2015). Turkish karst aquifers. *Environmental Earth Sciences*. https://doi.org/10.1007/s12665-015-4298-6.

Hamarat, S., Ülkenli, H., Türe, G., & Bayarı, C. S. (1998). Türkiye Kıyıları Aydıncık-Taşucu Deniz Mağaralarının Araştırılması. In *Türkiye'nin Kıyı ve Deniz Alanları 2. Ulusal Konferansı* (pp. 81–90). Ankara, Türkiye.

Haner, B., Yılmaz, A., Kürkçüoğlu, M. E., & Karadem, A. (2010). Mencilis (Bulak) Mağarasında Radon Seviyesi Ölçümleri. *Süleyman Demirel Üniversitesi Fen Bilimleri Enstitüsü Dergisi*. https://doi.org/10.19113/sdufbed.92370.

Helversen, O. V. (1989). New records of bats (Chiroptera) from Turkey. *Zoology in the Middle East, 3*(1), 5–18. https://doi.org/10.1080/09397140.1989.10637569.

Hill, C. A., & Forti, P. (1995). The classification of cave minerals and speleothems. *International Journal of Speleology 24*, 1–5.

Jones, M. D., Roberts, C. N., Leng, M. J., & Türkeş, M. (2006). A high-resolution late Holocene lake isotope record from Turkey and links to North Atlantic and monsoon climate. *Geology*. https://doi.org/10.1130/G22407.1.

Karacan, E., & Yılmaz, I. (1997). Collapse dolines in miocene gypsum: An example from SW Sivas (Turkey). *Environmental Geology*. https://doi.org/10.1007/s002540050125.

Klimchouk, A. B., Nazik, L., Bayari, C. S., Törk, K., & Kasjan, Y. (n. d.). *Kuzgun Cave and its Context: the first super-deep cave in the Aladaglar Massif, Turkey*. Retrieved from https://www.academia.edu/3144406/Kuzgun_Cave_and_its_Context_the_first_super-deep_cave_in_the_Aladaglar_Massif_Turkey.

Klimchouk, A., Bayari, S., Nazik, L., & Törk, K. (2006). Glacial destruction of cave systems in high mountains, with a special reference to the Aladaglar massif, Central Taurus, Turkey. *Acta Carsologica*.

Köhler, G., Renker, C., & Luis, P. (2004). A new cave population of Discoptila beroni Popov, 1975 (Insecta: Ensifera: Gryllidae) from southern Turkey, with some remarks on the genus. *Annalen des Naturhistorischen Museums in Wien. Serie B für Botanik und Zoologie*.

Kopar, İ. (2009). Aladağlar'da [Orta Toroslar (Yahyalı-Kayseri)] iki fosil mağara: Zindan-ı Köşk–1 ve Zindan-ı Köşk-2 mağaraları. *Türk Coğrafya Dergisi*. https://doi.org/10.17211/tcd.81508.

Külköylüoılu, O., Yavuzatmaca, M., Karacaoılu, D., & Telli, M. (2014). The aquatic diversity of ostracoda, phytoplankton and zooplankton from freshwater cave habitats in Turkey. *Cave and Karst Science*.

Kunt, B. K., Yağmur, A. E., Durmuş, H., & Anlaş, S. (2010). Checklist of the cave dwelling invertebrates (Animalia) of Turkey. *Biological Diversity and Conservation, 2*(3), 26–41.

Kunt, K. B., Özkütük, R. S., Elverici, M., Marusik, Y. M., & Karakaş, G. (2016). *Harpactea karaschkhan* sp. n., a new cave-dwelling blind spider species from the Mediterranean Region of Turkey. *Journal of Cave and Karst Studies*. https://doi.org/10.4311/2015LSC0106.

Laiz, L., Groth, I., Schumann, P., Zezza, F., Felske, A., Hermosin, B., et al. (2000). Microbiology of the stalactites fromGrotta dei Cervi, Porto Badisco, Italy. *International Microbiology, 3*, 25–30.

Legatzki, A., Ortiz, M., Neilson, J. W., Casavant, R. R., Palmer, M. W., & Rasmussen, C. et al (2012). Factors influencing observed variations in the structure of bacterial communities on calcite formations in Kartchner Caverns, AZ, USA. *Geomicrobiology Journal, 29*, 422–434.

Macalady, J. L., Dattagupta, S., Schaperdoth, I., Jones, D. S., Druschel, G. K., & Eastman, D. (2008). Niche differentiation among sulfur-oxidizing bacterial populations in cave waters, *ISME Journal: Multidisciplinary Journal of Microbial Ecology, 2*(6), 590–601.

Nazik, L., Törk, K., Güner, İ. N., Mengi, H., Özel, E., Aksoy, B., … Kara, A. (2001). Deniz kenarında bulunan polijenik bir mağara: Gilindire Mağarası (Aydıncık, İçel). *54. Türkiye Jeoloji Kurultayı*. Ankara.

Northup, D. E., & Lavoie, K. H. (2001). Geomicrobiology of Caves: a review, *Geomicrobiology Journal, 18*(3), 199–222.

Okay, A. (2008). Geology of Turkey: a synopsis. *Anschitt, 21*, 19–42.

Özbek, M. (2007). *Gammarus izmirensis* sp. nov., a new species of freshwater amphipod from Turkey (Amphipoda, Gammaridae). *Crustaceana*. https://doi.org/10.1163/156854007782605592.

Özbek, M. (2012). A new freshwater amphipod species, *Gammarus obruki* sp. nov., from Turkey (Amphipoda: Gammaridae). *Turkish Journal of Zoology*. https://doi.org/10.3906/zoo-1112-2.

Özbek, M. (2012). A new freshwater amphipod species, *Gammarus katagani* sp. nov., from turkey: (Amphipoda: Gammaridae). *Zoology in the Middle East*. https://doi.org/10.1080/09397140.2012.10648917.

Özbek, M., & Güloğlu, M. O. (2005). A new cave Amphipod from Turkey: *Gammarus ustaoglui* sp. nov. *Israel Journal of Zoology*. https://doi.org/10.1560/2WME-1HKT-QDD2-B3P9.

Özbek, M., Yurga, L., & Külköylüoğlu, O. (2013). *Gammarus baysali* sp. nov., a new freshwater amphipod species from Turkey (Amphipoda: Gammaridae). *Turkish Journal of Zoology*. https://doi.org/10.3906/zoo-1209-14.

Özdemir, A. (2015). Investigation of sinkholes spatial distribution using the weights of evidence method and GIS in the vicinity of Karapinar (Konya, Turkey). *Geomorphology*. https://doi.org/10.1016/j.geomorph.2015.04.034

Özdemir, A. (2016). Sinkhole susceptibility mapping using logistic regression in Karapınar (Konya, Turkey). *Bulletin of Engineering Geology and the Environment*. https://doi.org/10.1007/s10064-015-0778-x.

Özdemir, F., & Erkakan, F. (2014). The first new cave fish species *Cobitis damlae* (Teleostei: Cobitidae) from Turkey. *Hacettepe Journal of Biology and Chemistry*. https://doi.org/10.15671/hjbc.20144210863.

Özdemir, Y. (2005). Mencilis Mağarası. *Doğu Coğrafya Dergisi*.

Özkan, N. (2009). Chironomidae (Diptera) and Gammaridae (Amphipoda) Fauna in Dupnisa Cave (Sarpdere Village, Demirköy, Kırklareli). *E.Ü. Su Ürünleri Dergisi, 26*(1), 07–10.

Özşahin, E., & Kaymaz, Ç. (2014). Gilindire (Aynalıgöl) Mağarası'nın Turizm Potansiyeli (Aydıncık, Mersin). *Doğu Coğrafya Dergisi, 19* (1), 145–166.

Öztan, M., Baştanlar, Y., Varinlioglu, G., Hamarat, S., Ülkenli, H., Özyurt, N., & Bayari, C. S. (n.d.). *Kaş dolayı kıyı kuşağında karstlaşmanın gelişimi: Mivini ve Altuğ denizaltı mağaraları*. Retrieved from http://sat.metu.edu.tr/wp-content/uploads/2017/05/SBT2004-karstlasma.pdf.

Özyurt, N. N. (2008). Analysis of drivers governing temporal salinity and temperature variations in groundwater discharge from Altug Submarine Karst Cave (Kas-Turkey). *Environmental Geology*. https://doi.org/10.1007/s00254-007-0858-8.

Özyurt, N. N., & Bayarı, C. S. (2014). Helium isotopes as indicator of current hypogenic karst development in Taurids karst region, Turkey. In *Hypogene Cave Morphologies: Selected Papers and Abstracts of the Symposium Held February* (pp. 77–81).

Paksuz, S., & Özkan, B. (2012). The protection of the bat community in the Dupnisa Cave System, Turkey, following opening for tourism. *ORYX*. https://doi.org/10.1017/S0030605310001493.

Paksuz, S., Özkan, B., & Postawa, T. (2008). *Seasonal changes of cave-dwelling bat fauna, and their relationship with microclimate in Dupnisa Cave System (Turkish Thrace)*. Acta Zoologica Cracoviensia—Series A: Vertebrata. https://doi.org/10.3409/000000007783995435.

Ribera, C., Elverici, M., Kunt, K. B., & Özkütük, R. S. (2014). *Typhlonesticus gocmeni* sp. n., a new cave-dwelling blind spider species from the Aegean region of Turkey (Araneae, Nesticidae). *ZooKeys*. https://doi.org/10.3897/zookeys.419.5739.

Rowe, P. J., Mason, J. E., Andrews, J. E., Marca, A. D., Thomas, L., Van Calsteren, P., … Al-Omari, S. (2012). Speleothem isotopic evidence of winter rainfall variability in northeast Turkey between 77 and 6 ka. *Quaternary Science Reviews*. https://doi.org/10.1016/j.quascirev.2012.04.013.

Rowe, P. J., Wickens, L. B., Sahy, D., Marca, A. D., Peckover, E., Noble, S., … Andrews, J. E. (2020). Multi-proxy speleothem record of climate instability during the early last interglacial in southern Turkey. *Palaeogeography, Palaeoclimatology, Palaeoecology*. https://doi.org/10.1016/j.palaeo.2019.109422.

Schabereiter-Gurtner, C., Saiz-Jimenez, C., Pinar, G., Lubitz, W., & Rölleke, S. (2004). Phylogenetic diversity of bacteria associated with Paleolithic paintings and surrounding rock walls in two Spanish caves (Llonín and La Garma). *FEMS microbiology ecology*, *47*(2), 235–247.

Semenderoğlu, A. (2013). Fetrek Mağaraları (Vişneli-Kemalpaşa). *Marmara Coğrafya Dergisi*.

Semenderoğlu, A., & Aytaç, A. S. (2013). Kumalar Dağı'nda Psödokarstik bir mağara: Çakmaktepe Mağarası (Kumalar Dagi-Sand. *The Journal of Academic Social Science Studies*. https://doi.org/10.9761/jasss1024.

Solomon, S. B., Langroo, R., Lyons, R. G., & James, J. M. (1997). Radon exposure to tour guides in Australian show caves. *Environment International*. https://doi.org/10.1016/S0160-4120(96)00139-0.

Steiner, H. M., & Gaisler, J. (1994). On a collection of bats (Chiroptera) from NE Turkey and N Iran. *Acta Scientiarum Naturalium, Academiae Scientiarum Bohemicae Brno*.

Szerbin, P. (1996). Radon concentrations and exposure levels in Hungarian caves. *Health Physics*. https://doi.org/10.1097/00004032-199609000-00014.

Taşdelen, S. (2018). İnsuyu Mağarası'nın kuru gölleri ile Çine Ovası Akiferi arasındaki Hidrodinamik ilişki. *Ömer Halisdemir Üniversitesi Mühendislik Bilimleri Dergisi*. https://doi.org/10.28948/ngumuh.444795.

Taylan, M. S., Di Russo, C., Rampini, M., & Cobolli, M. (2011). The Dolichopodainae and Troglophilinae cave crickets of Turkey: An update of taxonomy and geographic distribution (Orthoptera, Rhaphidophoridae). *Zootaxa*. https://doi.org/10.11646/zootaxa.2829.1.3.

Taylan, M., Yılmazer, N., & Şirin, D. (2020). A comparative study on temperature and relative humidity data of three caves in different climatic regions of Turkey, with notes on the distribution of Anatolian Cave crickets (Insecta, Orthoptera, Rhaphidophoridae). *Journal of the Entomological Research Society, 22*, 53–73.

Tok, E. (2017). The bacterial and archaeal biodiversity of Insuyu cave/İnsuyu mağarasının bakteriyel ve arkeal biyoçeşitliliği. İstanbul Teknik Üniversitesi/Avrasya Yerbilimleri Enstitüsü.

Ünal-İmer, E., Shulmeister, J., Zhao, J. X., Tonguç, I., Feng, Y. X., Duc Nguyen, A., & Yüce, G. (2015). An 80 kyr-long continuous speleothem record from Dim Cave, SW Turkey with paleoclimatic implications for the Eastern Mediterranean. *Scientific Reports*, *5*, 13560. https://doi.org/10.1038/srep13560; https://www.nature.com/articles/srep13560%23supplementary-information.

Ünal-İmer, E., Shulmeister, J., Zhao, J. X., Uysal, I. T., & Feng, Y. X. (2016). High-resolution trace element and stable/radiogenic isotope profiles of late Pleistocene to Holocene speleothems from Dim Cave, SW Turkey. *Palaeogeography, Palaeoclimatology, Palaeoecology*. https://doi.org/10.1016/j.palaeo.2016.04.015.

Uzun, A., Zeybek, H. İ., Yılmaz, C., & Bahadır, M. (2015). Aksu çayı havzasında traverten mağaraları, Giresun. *Marmara Coğrafya Dergisi*. https://doi.org/10.14781/mcd.95521.

Wickens, L. B. (2013). *Geochemistry and Petrography Of speleothems from Turkey and Iran: Palaeoclimate and Diagenesis*. UK: Universityof East Anglia.

Yamaç, M., Işik, K., & Şahin, N. (2011). Numerical classification of streptomycetes isolated from karstic caves in Turkey. *Turkish Journal of Biology*. https://doi.org/10.3906/biy-0911-185.

Yang, H., Johnson, K. R., Griffiths, M. L., & Yoshimura, K. (2016). Interannual controls on oxygen isotope variability in Asian monsoon precipitation and implications for paleoclimate reconstructions. *Journal of Geophysical Research*. https://doi.org/10.1002/2015JD024683.

Yılmaz, E., & Çiçek, İ. (2016). Thornthwaite climate classification of TurkeyTürkiye Thornthwaite iklim sınıflandırması. *Journal of Human Sciences, 13*, 3973. https://doi.org/10.14687/jhs.v13i3.3994.

Yücel, S., & Yamaç, M. (2010). Selection of streptomyces isolates from Turkish karstic caves against antibiotic resistant microorganisms. *Pakistan Journal of Pharmaceutical Sciences*.

Check for updates

45 Interesting Caves of Turkey

4

Abstract

In addition to archaeologically and structurally important ones, the deepest and longest caves of Turkey are explained in this chapter. These 45 caves are in different regions of Turkey and naturally have different geological properties. Even though three of the four deepest caves of Turkey are almost at the same spot, all the longest caves reside in different regions and within very different geological formations. In order to be understood more easily, comprehensive geological and hydrogeological explanations are given for all the caves explained in this chapter.

4.1 Introduction

Approximately 40% of Turkey's total area of 783,562 km^2 consists of rocks such as limestone, dolomite, and gypsum which are suitable for karstification. With a rough estimation, it is predicted as there are thousands of caves in these karst regions reaching a total area of approximately 300,000 km^2 (Nazik, 2004; Nazik, Poyraz, & Karabıyıkoğlu, 2019). However, due to the speleological researches which have started on a late date as the 1960s, only a small number of these caves have been explored and surveyed. TAY Project Cave Inventory of Turkey (www.tayproject.org) carried out based on the published documents, includes 3,050 caves for the present (Gürcan et al., 2006). This number is so much less than the number of caves in numerous countries which are smaller than Turkey. Just to give an example, in countries such as Italy, Romania, Switzerland, and Slovenia, it is known that there are more than 10,000 inventoried caves, and this number is 8,500 in Greece.

The caves explored and researched in Turkey until now are predominantly centered in the Taurus Mountains, West Anatolia, and Western Black Sea Regions. The number of caves in the Eastern Black Sea and Central Anatolian regions, which are not geomorphologically suitable for the formation of caves, is limited. On the contrary, the number of researches carried out for the caves in the Southeastern Region, outcropped with limestone on quite wide areas, is not sufficient.

Though the number of researches carried out is limited, the caves explored in Turkey until today can be classified in different categories. One of these categories can be as the caves in regions differing in terms of the geographical characteristics (Nazik & Mulazimoglu, 2007). The caves in Turkey show different morphological characteristics in different regions depending on the geological and meteorological differences. For example, there are serious structural differences between the caves in the Mediterranean Region and caves in the Black Sea Region. It is known that the Taurus Mountains, which is in the south of the Anatolia with a length of more than 1000 km from east to west, has a limestone thickness of 4000 m. This mountain range, which is constantly bending for millions of years with the tectonic movements, includes numerous deep sinkholes, some of which are with depths more than −1000 m. On the other hand, nearly all the caves explored until now in the Black Sea Region are horizontal due to the slight thickness of limestone and low tectonic movements and the deepest cave of this region is less than −300 m. Similarly, the caves in the Western Anatolian Region and Marmara Region mostly show a horizontal development (Nazik & Tuncer, 2010).

Apart from these structural differences mentioned above, a comprehensive classification concerning the speleogenesis or geomorphological structures of the caves in Turkey has not been made yet. Except for the limestone and dolomite rocks, the third most common group of rocks in which caves have been formed is the gypsum rocks, and most of these caves are specifically located in Sivas area. Though there are numerous caves explored in gypsum rocks in this area, all of them are quite small. Except for the mixed formations, the only big cave formed entirely in the conglomerate rocks is Tilkiler Cave. Apart from this cave, a few more small

A. Yamaç et al., *Caves and Karst of Turkey—Vol. 1*, Cave and Karst Systems of the World,
https://doi.org/10.1007/978-3-030-65501-3_4

conglomerate caves have been explored and surveyed in Antalya–Köprüçay Canyon on a wide area outcropped with the conglomerate rocks (Değirmenci, Bayarı, Denizman, & Kurttas, 1994). Apart from these four groups of rocks mentioned above, though there are volcanic areas especially in the Central Anatolia and Eastern Anatolia regions, besides small examples, there is no important volcanic cave explored until now. Hypogenic cave researches have accelerated in recent years. Regarding this issue, there are certain studies of different experts about Insuyu Cave and dolines in Obruk Plateau–Konya (Bayarı, Ozyurt, & Pekkan, 2009; Bayarı et al., 2017).

In addition to all these different cave categories, the caves in Turkey can be subject to a different classification in terms of the "human usage". It is known that a certain part of the thousands of caves in the Anatolia has been used by the inhabitants of these lands throughout history and is still used. Even besides the archaeological periods, this usage arising from different socioeconomical purposes during different periods and in different regions varies on a wide range from food storage to cheese production and from a barn to worshiping and/or hiding. This fact, which is archaeological in many other countries, still exists in Turkey, and even today there are caves still used for those purposes. Even apart from the thousands of artificial caves dig in the volcanic rocks in the Cappadocia Region located in the Mid-Anatolian Volcanic Province and some of which are still used, it is a rather ordinary situation for the cavers to encounter piles of cheese or lemon or a small herd of goats in a natural cave entered for exploration. Similarly, though the Southeastern Anatolian Region, which is almost entirely limestone, is closed for research due to the conflicts faced for approximately 35 years, it is known that numerous caves in this region are still used as hiding places.

This chapter contains 45 examples chosen among the caves of Turkey explored and surveyed on different dates. While preparing the list, we have tried to choose caves from different regions with different formation processes and different geomorphological characteristics as to show a general sample about the caves in Turkey. While some of these caves are very impressive in terms of speleothems, some others are vertical shafts without almost any cave formation. In the list, there are the longest caves of Turkey such as Pınargözü, İnsuyu, and Tilkiler and the deepest sinkholes such as Peynirlikönü, Kuzgun, and Morca. We have also included the most important caves in terms of the results of the archaeological excavations such as Karain, Öküzini, and Yarımburgaz and few interesting show caves such as Dim and Oylat.

Though there are rather allegoric and meaningful Turkish names of certain caves such as Cennet and Cehennem (Heaven and Hell) and Kırkgözler (Forty Springs), we have preserved the Turkish names of all caves upon considering that the usage of the English translation of these names will detract the cave referred from the original context, and moreover we will have to translate most of the cave names in such a case. Likewise, you will realize that though certain caves are typical sinkholes, their names are stated as "caves". The reason for this is that we do not want to change the local names or the names given to these caves by the explorers.

We would like to inform that, though the entire list below concerning the longest and deepest caves of Turkey is from a published source and up to date, it might change constantly in light of the researches carried out each year.

Longest caves of Turkey (2020)

Cave	Province	Length (m)
Pınargözü Cave	Yenişarbademli, Isparta	8500
Insuyu Cave	Burdur	8350
Tilkiler Cave	Manavgat, Antalya	6818
Kızılelma Cave	Zonguldak	6630
Yaylacık–Inilti Pazarı System	Gündoğmuş, Antalya	5929
Bulak Mencilis Cave	Karabük, Safranbolu	5250
Altınbeşik Cave	Akseki, Ürünlü, Antalya	5119
Ayvaini Cave	Ayvaköy, Bursa	4866
Ikigöz Cave	Çatalca, Istanbul	4816
Morca Sinkhole	Anamur, Içel	4068
Yazören Cave	Yazören, Balıkesir	3554
Çukurpınar Sinkhole	Anamur, Içel	3350
Gökgöl Cave	Erçek, Zonguldak	3350
Kuzgun Sinkhole	Niğde	3187
Dupnisa Cave	Sarpdere, Kırklareli	3150
Peynirlikönü Sinkhole	Anamur, Içel	3118
Düdenağzı Sinkhole	Başyayla, Karaman	2528
Susuz Cave	Seydişehir, Konya	2303
Tınaztepe Caves	Seydişehir, Konya	2195
Kızılin Cave	Burdur	2176
Saçayağı Cave	Gazipaşa, Antalya	2125

Deepest Caves of Turkey (2020)

Cave	Province	Depth
Peynirlikönü Sinkhole	Anamur, Içel	−1429
Kuzgun Sinkhole	Niğde	−1400
Morca Sinkhole	Anamur, Içel	−1210
Çukurpınar Sinkhole	Anamur, Içel	−1196
Kuyukule Sinkhole	Dedegöl, Isparta	−832
Keş Sinkhole	Kahramanmaraş	−728
Subatağı Sinkhole	Yahyalı, Kayseri	−643
Sütlük Sinkhole	Pozantı, Adana	−640
Düdenağzı Sinkhole	Başyayla, Karaman	−612
Çem Sinkhole	Tomarza, Kayseri	−605
Yılanlıyurt Sinkhole	Aladağ	−603
Yaylacık – Inilti Pazarı System	Gündoğmuş, Antalya	−595
Kocadağ Sinkhole	Anasultan, Kütahya	−458
Pınargözü Cave	Yenişarbademli, Isparta	+440
Düdenyayla Sinkhole	Beyşehir, Konya	−416
Atlılar Sinkhole	Gözne, Içel	−410
Çamlıköy Sinkhole	Pozantı, Adana	−379
Macar Sinkhole	Gazipaşa, Antalya	−356
Bucakalan Sinkhole	Akseki, Antalya	−345
Ölü Köpek Sinkhole	Akseki, Cevizli, Antalya	−340
Düdencik Sinkhole	Akseki, Cevizli, Antalya	−330

4.2 Interesting Caves of Turkey

4.2.1 Altınbeşik Cave

Ürünlü–Antalya
Total Length: 4538 m
Total Depth: +134 m

Altınbeşik Cave is an outlet cave running into the valley of Manavgat River located in 2 km east of Ürünlü Village in the province of İbradi. The entrance is 150 m above the riverbed of Manavgat River. In the region, generally, the Cretaceous and Jurassic–Cretaceous aged limestone formations are observed, and Altınbeşik Cave is formed in the curves formed by the Upper Cretaceous aged limestone. The basic geological structure of the region is shaped as strong curves occurring during the Tertiary period and strong vertical inclines at the end of this period. The incline of the area has mostly occurred as a result of the epeirogenic movements. However, it is also possible to state that the faulting and Alpine Orogeny have an important part in the shaping of the land. The area under the effect of the Alpine Orogeny shows a curved and faulted structure. Concerning the current shaping of the region, in addition to the lithological structure, factors such as tectonism, karstification, and stream erosion have been effective as well (Akay & Uysal, 1988). These factors have served to the shaping of the geomorphological structure of the area such as various valley types, caves, dolines, erosion surfaces, and debris cones. The sources of Manavgat River, the most important stream of the region, come from the surroundings of Akseki and İbradı Plateaus and 2/3 of the stream is from the karstic sources (Kaya, Simsek, & Akis, 2015). One of these karst sources is Altınbeşik Cave and is located at the outlet of a very long and huge underground hydrological system. This huge system draining the waters of Kızılova, Kembos, and Söbüova poljes and the sinkholes located at the edge borders of these poljes end in Altınbeşik Cave (Aygen, 1984).

The entrance of the cave is in the form of a massive lake (Fig. 4.1). To access the first sump of the cave, it is necessary to go around seven lakes (Lake I–VII) in total and climb several vertical inclines. The most significant vertical incline is a 40 m high wall covered with speleothems above the lake at the entrance (Lake I). There is a deep and wide passage through which water flows in the springtime when the level of water is high. The dry fossil upper level of the cave is accessible only after the first 500 m of the cave. The shape of this upper level of the cave is similar to the main passage which is 30 m below. The speleothem structure is extensive only in the entrance and dry fossil upper level of the cave.

Altınbeşik Cave was first explored by SCP in 1966, and its first map was drawn (C. Chabert, 1966). Altınbeşik Cave was re-explored and remapped three more times in the later years. Due to various tracer tests carried out during the preparations for the construction of Oymapınar Dam, the connection between Altınbeşik Cave and sinkholes in the

Fig. 4.1 First lake of Altınbeşik Cave (*Photo* R. Straub)

southern part of Kembos Plain has been proven. The direct distance between the sinkholes and cave is more than 33 km. This fact has fascinated numerous speleologists and explorations have started to be carried out by several teams (Müller, 1975; Schmitt, 1976, 1980), and during these explorations, some dry galleries have been mapped. During a joint expedition carried out by the Americans, Germans, and Turks in 1986, the first siphon was dived for the first time (Hardcastle & Schmitt, 1987). Geospeleos of the Czech Speleological Society has carried out numerous expeditions during 1992, 1995, and 1997 to Altınbeşik Cave and its surroundings. During the expedition of Geospeleos in 1992, first siphon was dived again and a spacious dome has been documented behind this siphon which after 300 m led to another siphon. The second siphon was shallow and only a few meters long. The cave continues after the vertical 40 m high wall behind the second siphon. The passage after this wall is not as wide and deep as the one in front of the second siphon, and the area has a width of 10–20 m. During the next expedition, a new and massive passage has been explored with a width of 30 m and a height of 12 m. This passage continues all the way to the lake, above which there is a chimney. In front of the lake, the cave sharply turns to the east and swiftly descents by forming large curves. A tiny flow appears in the lake and afterward flows through the passage deeper in the cave. In the next zones, the passage continues below a pair of chimneys, the ceilings of which cannot be seen. At the end, which is 200 m away from the lake, the passage merges with the sump, and this third sump is named Nesvik (Jäger & Janoušek, 2013; Janousek, 2018).

4.2.2 Ayvaini Cave

Doğanalan–Bursa
Total Length: 4866 m
Total Depth: −80 m

Ayvaini Cave, located near Doğanalan Village in the province of Bursa, is the longest traverse cave of Turkey explored until now with a length of 4866 m (Fig. 4.2). It is formed in the Jurassic–Lower Cretaceous neritic limestone. The thickness of this limestone, located on the impermeable blocks of the Triassic aged Karakaya Formation and generally lined in the directions of northeast and southwest, is around 100–150 m. This limestone block is discordant among the Quaternary alluvium deposits on the north and Triassic clastic and carbonate rocks on the south. This Jurassic–Lower Cretaceous neritic limestone, giving its characteristics to the karst of the region, has the stratigraphical and structural characteristics suitable for karstification and cave formation. The impermeable blocks of

Fig. 4.2 Ayvaini Cave (*Photo* M. Albukrek)

Karakaya Formation, just located under this formation, is at the karst base level for the limestone. The plateau around Ayvaini Cave is formed of shapes belonging to the Plio-Quaternary relief system. Especially, the Pliocene period has been critically determinative in terms of the morphology of the region. The pieces of erosion surfaces, decomposed karst areas, fluvio-karstic uvala, hanging valleys, caves, and dolines found on heights between 400 and 600 m are the most characteristic shapes of the Pliocene (Nazik, Törk, & Özel, 1997).

Ayvaini Cave, located on the northeast side of the plateau 350–700 m higher than Uluabat Lake, is a huge underground system collecting the surface and underground waters of a wide area. Ayvaini Cave, without any side branch, consists of a single main gallery between a sinkhole, extending in the direction of southwest and northeast in parallel with the limestone block and a spring. The sinkhole, where Karadonlu Stream disappears by forming a 10 m waterfall, can be easily located along the course of the water. Though the level of water increases in the spring, there are plenty of lakes in the cave that can be passed. After the chamber right below the descent, the main gallery starts which continues with an incline of 1–2% and width between 4 and 10 m. There are many chimneys considered to carry the water from the surface dolines, and they also have large travertine depots underneath. The stalactites and dripstone pools are particularly among the interesting structures. On the contrary, since the cave is active, there is almost no stalagmite development. Though formations similar to chimney have been determined on the ceiling of the cave, no result has been achieved from the doline researches carried out on the surface between the sinkhole and spring, and no new connection to the cave has been noticed.

The first exploration in Ayvaini Cave was carried out in 1952 by K. Lindberg, a Swedish biospelelogist who conducted the first speleological surveys in Turkey (Lindberg,

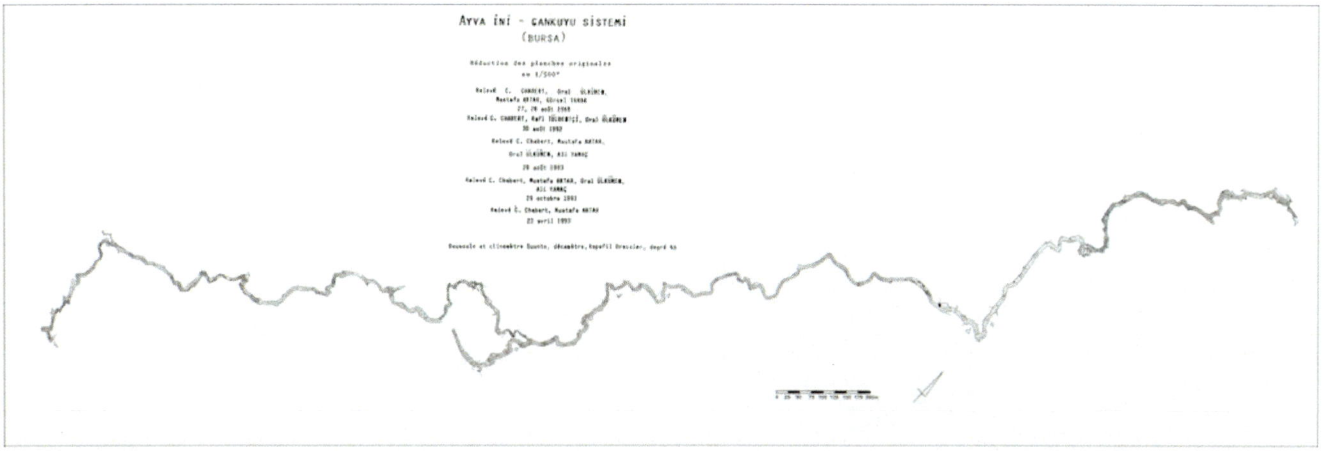

Fig. 4.3 Map of Ayvaini Cave compiled from the 40 sheets of unpublished atlas of Claude Chabert surveyed during 1988–1993

1952a, 1952b). After this exploration, in 1970, the Spanish cavers did the first traverse of the cave and draw a rough map (Masriera, 1971; Masriera & Martorell, 1971). It has been stated that during the research carried out by MTA, a small colony of bats was seen in the entrance zone of the cave though the species were not indicated (Nazik et al., 1997). Apart from all these researches, the most interesting study carried out in Ayvaini Cave is the 1/500 scale map of the cave started in 1988 and finalized in 1993 by Claude Chabert, a member of SCP (Fig. 4.3). It is the longest cave map of Turkey with a total length of 10 m and contains incredible details about Ayvaini which has a total length of 4866 m (Yamaç, 2013b).

4.2.3 Balatini Cave

Çamlık–Derebucak–Konya
Total Length: 1768 m
Total Depth: −32 m

Balatini Cave is located within the boundaries of Çamlık Village in the district of Derebucak in Konya in the mid-region of the Taurus Mountains. It is formed in the Jurassic–Cretaceous limestone of Beyşehir–Hoyran Nappes. Though the limestone of this region is very suitable for karstification, due to the slight thickness and impermeable Permian schist underneath, the caves have developed horizontally. Balat uvala, housing the cave on its edge, is formed by karstification, decomposition of the Pliocene relief system. The decomposition has resulted from the Post-Pliocene tectonic movements. Balatini Cave drains the waters of this uvala, which is a closed basin, to underground and drains out in a short distance. The cave has five entrances and includes three interconnected main branches. The two entrances on the uvala side drain the water, and three different spring mouths drain out from the slope facing Uzunsu River alongside a valley. Balatini is formed of two different levels on top of each other. The bottom of the fossil branch, compromising the upper level, is entirely covered with clay while in the water-carrying gallery at the lower level, and it is possible to walk through the gallery when there is not much water. The main gallery consists of two branches. Initially consisting of a single gallery, this active/semi-active gallery is divided into two in the midway. Between these two, the one on the left side is the active branch. With water flowing even in the most rainless season, it is the narrowest gallery of the cave with the lowest roof. The second branch of the main gallery, on the other hand, leads to the sinkhole entrance in the uvala. Three giant whirlpools, only one of which is deeper than 5 m, can be passed by the traverse technique or by boat. There are numerous calcite structures especially in the fossil branch and many lakes in the other galleries. Among these lakes, the depth of which varies between 0.5 and 2 m lies some islets of pebble and sand. Hydrologically, Balatini has three different zones: The uppermost gallery is completely at the zone; the second gallery following the second and lower entrance at the uvala has semi-phreatic specification, and the lower part of that second gallery is completely phreatic (Nazik, Güldali, Tüfekçi, Beydeş, & Aksoy, 1993).

The cave is probably much older than the other caves in the area and has changed considerably. The original underground watercourse followed a generally southwesterly route, similar to that of today but at a higher level, and today is in the fossil stage and completely devoid of any water. The passage size is generally large up to a width of 18 m and a height of 12 m and contains only moderate calcite deposition (Fig. 4.4).

Only the southernmost zones of this passage remain intact as the present-day river has taken a completely different course at that point. Elsewhere, the combination of collapse

Fig. 4.4 Balatini Cave (*Photo* B. Langford)

and downcutting has left only the remnants of the old passage visible high in the roof of the current watercourse. The present-day tributary stream passage is quite immature and is a relatively recent intrusion into the main river passage. The meandering rift joining the fossil passage some 160 m from its downstream end is almost certainly the original passage carrying the tributary stream The floor of this rift is several meters lower than that of the fossil passage indicating perhaps that it was active long after the main river had cut its lower course, and this would explain the collapse of the fossil passage joining it with the present-day river passage at that point. The upstream entrance is almost certainly the original entrance but the river would have turned sharp right just inside the cave, along a large passage which is now choked with the sediments. The very high ceiling in the main river passage at its northern end indicates where downcutting has taken place from the original passage above. There is some evidence to suggest that the upper passages were totally underwater at times and a much more comprehensive study would be required to discover the relationship between the changes which took place in the cave and downcutting of Uzunsu Valley outside (Stratford, 1991; Stratford & Crabbe, 1992).

Balatini Cave has been surveyed by Swindon Speleological Society (Fig. 4.5) and MTA in 1992 separately. According to the report of MTA, the potsherds and cultural deposits revealed on the upper fossil level are indicators of the archaeological significance of the cave (Nazik et al., 1993).

4.2.4 Bayındır Cave

Bayındır–Büyükorhan–Bursa
Total Length: 266 m
Total Depth: −31 m

Bayındır Cave is located in the South Marmara region, and it was developed within Triassic aged marbles in the Tavşanlı Zone. Those are usually gray, dirty white, bluish, coarse crystalline marbles and occasionally contain mica levels. Those marbles cut by the granites which are tectonically overlain by Yayla Melange. In this region carbonate rocks which is suitable for karstification cover large areas. Permo-Triassic limestones and marbles and also Jurassic limestones are the main formations of this region. Triassic marbles at the west of "Çakıllı Tepe" are hydrologically controlled by Karanlıkdere and Çaydere creeks (Nazik et al., 1997). Located in this area, the general direction of the Bayındır Cave is northwest–southeast. It has an average width of 5.5 m and a ceiling height of 5.6 m. Although the base slope is in areas where the upper ground is higher than 15°, and this is the case for a limited distance. The entrance width is 2 m, and the height is 2.5 m. The cave starts with a sloping floor, and this section continues about 60 m then reaches an area filled with cave deposits. In the first hall, an intense amount of guano is observed. On the right side of the first hall, there is another small hall with a total length of 28 m and a width of 5 m, rich in cave deposits. With the first hall, the richness of the Bayındır Cave is in terms of cave deposits began to increase. At the end of the hall, the second hall is reached after a steep descent of about 4 m formed by the cave columns. Various dripstones are also observed in this section. The third hall of the cave reached through a passage with a width of 1 m and a high of 0.5 m and covered with dripstones. In the third hall of the Bayındır Cave, cave deposits are observed as well as large and small rock blocks. At the end of these blocky areas, the most ornamented fourth and last hall of the cave is reached through a passage (Fig. 4.6). This hall is 60 m long and has an average width of 6 m. The ceiling height of this section is 2 m at the entrance and reaches up to 10 m (Törk, Savas, Yeleser, & Kahraman, 2014).

4.2.5 Birkleyn Caves

Lice–Diyarbakır

Birkleyn Cave system, located near Lice, 100 km northeast of Diyarbakır in the southeastern region of Turkey contributes to the importance of karstification and the karstic formation features. This system, consisting of four different caves, has significant importance both in geomorphological and archaeological aspects. The area of research corresponds to a thrust belt on the southeastern Taurus Mountains. The formation of Birkleyn Cave system and its geomorphological characteristics give some clues about the collision of two lithospheric plates called the Anatolian microplate and

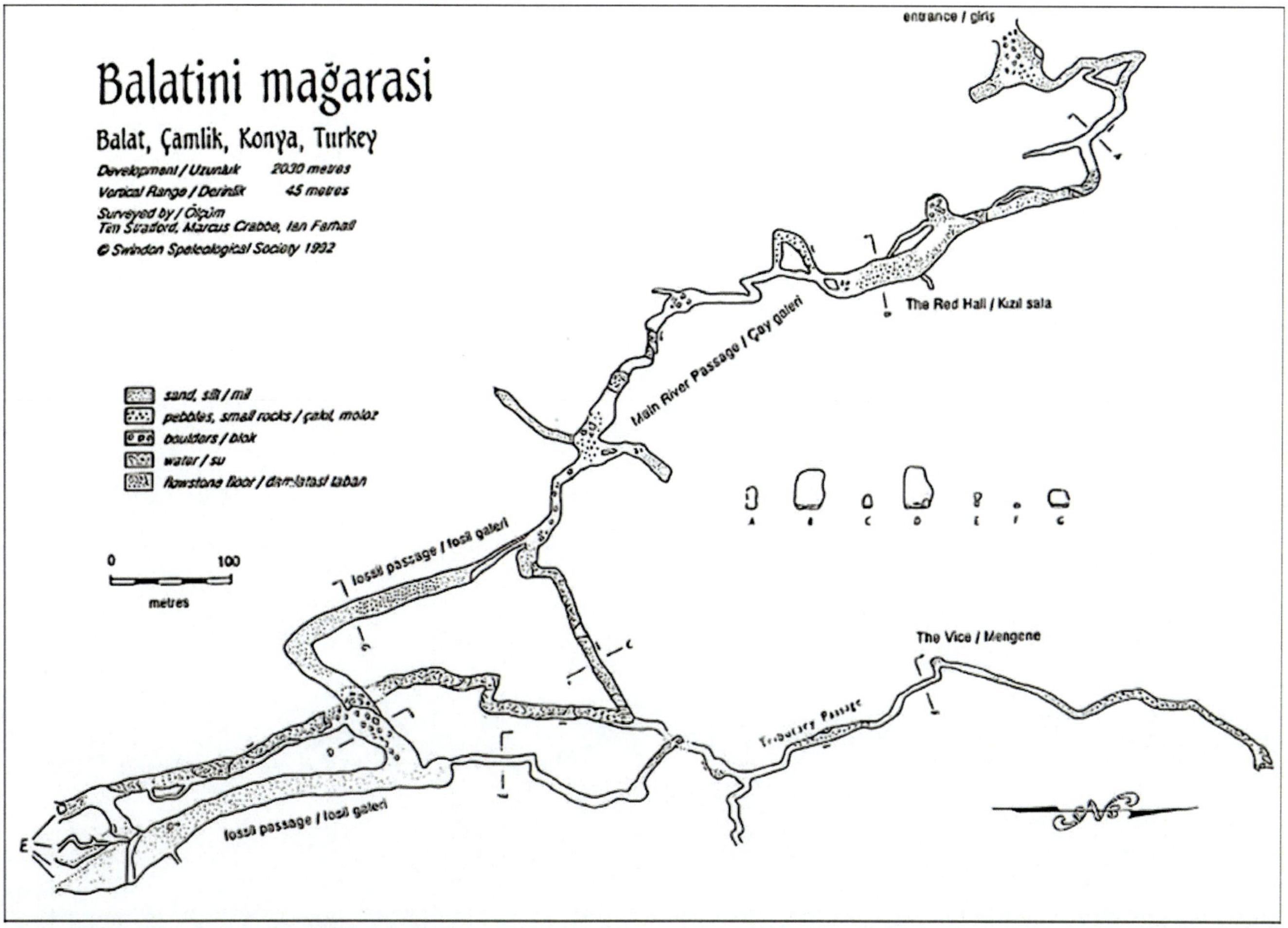

Fig. 4.5 Map of Balatini Cave (Stratford & Crabbe, 1992)

Fig. 4.6 Last section of Bayındır Cave (*Photo* K. Törk)

Arabian plate. In the region, generally, the Cretaceous and Eocene aged volcano-sedimentary rocks and weathered ophiolites are dominant. However, the Korha Mountain block is a tectonic window, which has occurred with erosion and tectonism and entirely has the structure of the Eocene aged calcareous (I. Atalay, Karadogan, & Yıldırım, 2010). Dibni Su, one of the two important branches of Tigris River, takes its source from this area, and this river, established on the syncline on the Korha Mountain, has shifted from the surface to underground in accordance with the incline of the area and rejuvenation of the karstification. Due to this branch feeding Tigris River and flowing underground, this cave system was assumed as the source of Tigris River for centuries. Until the Quaternary period, the watercourse of Dibni Su was quite different. Initially, Dibni Su was flowing on the syncline at the current high levels in the direction of northeast and southwest after passing the northwestern edge of the Korha Mountain. This valley was blocked with a collapse occurring in the Quaternary and the stream searched for new ways out. First, the stream found its way out from the caves within the nummulite limestone on the right bank. However, as a result of the incline of the area based on the early Quaternary tectonism, settled in the 1st cave, where it still flows today and found its way out. Birkleyn Caves have been formed around this valley opened at the northwest edge of the Korha Mountain anticlinal in the direction of northwest and southeast. Dibni Su flows out of the 1st cave on the faulted southwestern slope of the anticlinal and continues to flow from the outlet of the valley by connecting to the anticlinal again with a natural bridge (I. Atalay et al., 2010). The length of the 1st cave is 870 m (Schachner, 2005) and continues along the Korha Mountain slope with zigzags and forms the southern part of the canyon (Fig. 4.7).

In the last part of the 1st cave, in the gap formed due to the collapse, there are in total three inscriptions and two reliefs. The Assyrian Kings Tiglath Pileser I (1114–1076 BC) (Fig. 4.8) and Shalmaneser III (858–824 BC) carved "images of kingship" and accompanying royal inscriptions on this wall (Harmansah, 2007). All those inscriptions are located at a height of 4–5 m above the current base level. On the western, the inscription and relief of Tiglath Pileser I are found and the relief and one inscription of Shalmaneser III

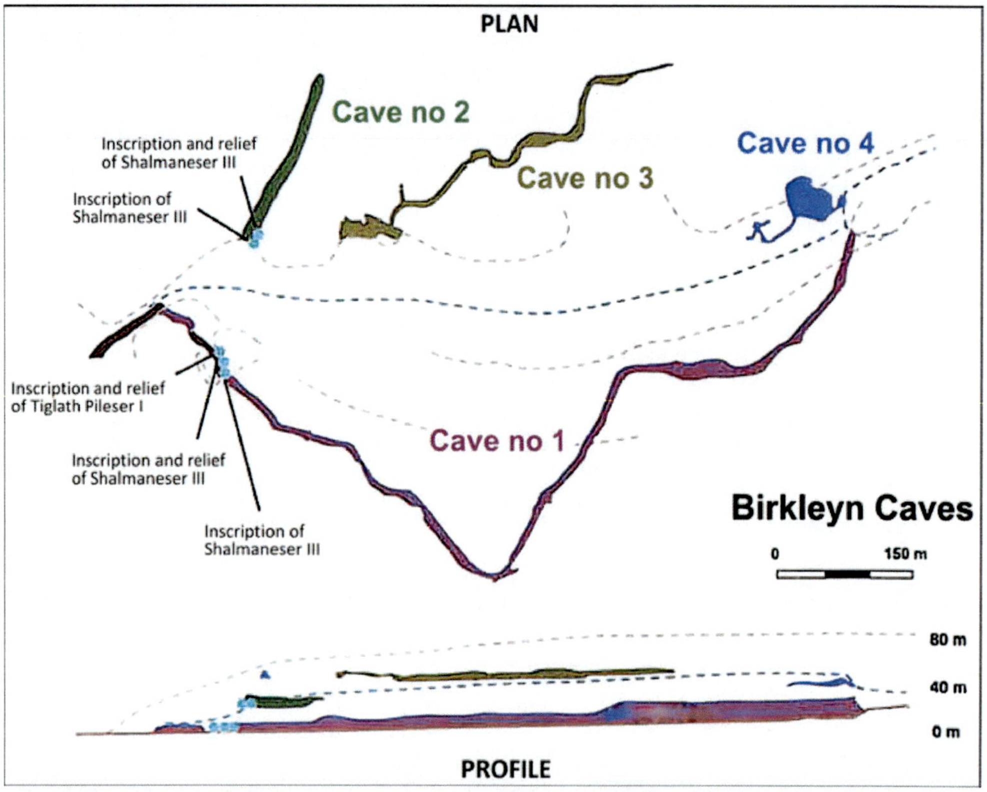

Fig. 4.7 Plan of Birkleyn Caves, adapted from (Schachner, 2009)

Fig. 4.8 Inscription and relief of Tiglath Pileser I on the wall of Birkleyn Cave no. 1 (*Photo* A. Ersöz)

and approximately in 10 m another inscription of the same king are seen at the same place as well. The cuneiform inscription of the Assyrian King Tiglath Pileser I at this place mentions that he visited the cave on three separate occasions. This is the world's first dated inscription in or about a cave.

The entrance of the second cave is located at a height approximately 40–45 m above the base level of the valley on the northern slopes of the water gap. This is located at a place 50 m higher than the first cave. This cave, the length of which is 250 m with a width of 30 m and height of 25 m at the widest point, is the widest and most used cave of Birkleyn Caves. There are remains in the cave belonging to different periods from the Iron Age to the Byzantine Period. On the exterior of the second cave, on the east rocky side of the platform on which a huge natural rock cluster is located, there are two more inscriptions and another relief. All these second group monuments belong to Shalmaneser III. The pose of the king is similar to the one in the 1st cave. The right hand is raised for praying, and the left hand is on the hilt. He is wearing a long dress, and the upper end of the dress is tied on the chest. The inscriptions are on the right side of the king and different from the first cave, and they are behind the king. Despite their differences, in all these five inscriptions mentioned, the conquests of these two kings and their victories in the region are described (Schachner, 2005, 2009).

The third cave, with a bit narrower entrance, is located at a height 60–70 m above the base level of the valley and approximately 50 m above the second cave, again on the right slope of the valley. After the narrow entrance and a small chamber, the quite long main gallery of the cave is entered. The gallery is quite rich in terms of stalactites, stalagmites, columns, and travertine. The length of this cave is approximately 450 m. Fourth cave is located on the eastern of the canyon and northern edge of the slope. This cave consists of passages enabling to pass to the interior area of the mountain and two big corridors connecting to the slots. In this cave, few bats (Myotis myotis, Eptesicus serotinus) have been reported (Benda & Horacek, 1998).

Birkleyn Caves have drawn the attention of numerous researchers for many years due to the Assyrian inscriptions in the caves. In 1862, J. G. Taylor visited the area and presented a description of the inscriptions for the first time (Taylor, 1865). The study carried out by Lehmann-Haupt in 1899 is accepted as the first comprehensive archaeological research, and he was the first Assyriologist to publish the inscriptions and reliefs (Lehmann-Haupt, 1901, 1910). After numerous different publications and brief archaeological research carried out in 1986 (Russel, 1986), the comprehensive studies have started in 2004 under the presidency of Andreas Schachner from the University of München. Apart from these researches, Birkleyn Caves have been explored and researched also by different cavers since 1976 (Halliday, 2001; Halliday & Shaw, 1995; Kusch, 1993a, 1993b; Timing, 1984; Waltham, 1976).

4.2.6 Bulak Mencilis Cave

Bulak–Karabük
Total Length: 5250 m
Total Depth: −4 m +162 m

Bulak Mencilis Cave is located on the northwest of Bulak Village between Karabük and Safranbolu. The Cretaceous limestone has outcropped as a quite huge profile on this part of the Western Black Sea Region in the direction of southwest and northeast (MTA, 1964). Both the north and south of this limestone band have been limited with the Eocene flysch, and the karstification is shallow because of the underlying impervious rocks. Bulak Mencilis Cave formed in this Cretaceous limestone consists of two cave development floors formed in different periods. The longest and youngest zone of the cave, with three entrances to the relevant floors, is the ground floor. In this zone hydrologically active, there is a big underground stream, the flow rate of which changes seasonally between 0.55 and 2.20 m^3/s with a significant flow during summer and winter and numerous lakes in different sizes. Mencilis karst spring has recharged directly by the limestone developing the karstic channel system (Törk, 1995) and that is the reason for the short resurgence time of water in Mencilis karst spring. Bulak Mencilis Cave, together with the other resurgence cave of the region, namely Hızar Cave, meets the main part of the freshwater requirement of Karabük and Safranbolu provinces. The karstic springs of Mencilis and Hizar drain most of the rain from the basin. Mencilis spring has drained the developed karstic system, and Hizar spring has also drained by the small fractured systems (Törk, 1994). Bulak Mencilis Cave has two entrances. The first one is through a mouth with water outflow but it ends with a siphon after 30 m This branch, after five consecutive siphons, joins to the main fossil gallery. The second entrance is on the left side of the hill behind the active outflow. This second entrance 100 m above the first spring entrance reaches the fossil gallery, and in this gallery, there are travertine lakes, stalactites, and stalagmites. This zone, covered with various dripstones, connects with the lower galley with five siphons. This upper floor, which is the first formed zone of the cave, is entirely in the vadose zone today, and the first 310 m of this zone has been opened to tourism in 2003. After this connection point, the cave is in the phreatic zone and is a semi-active single gallery. Two hundred m after this point, there is the sixth siphon. Between the sixth and seventh siphons, there is a 1500 m dry gallery. The researches about Bulak Mencilis

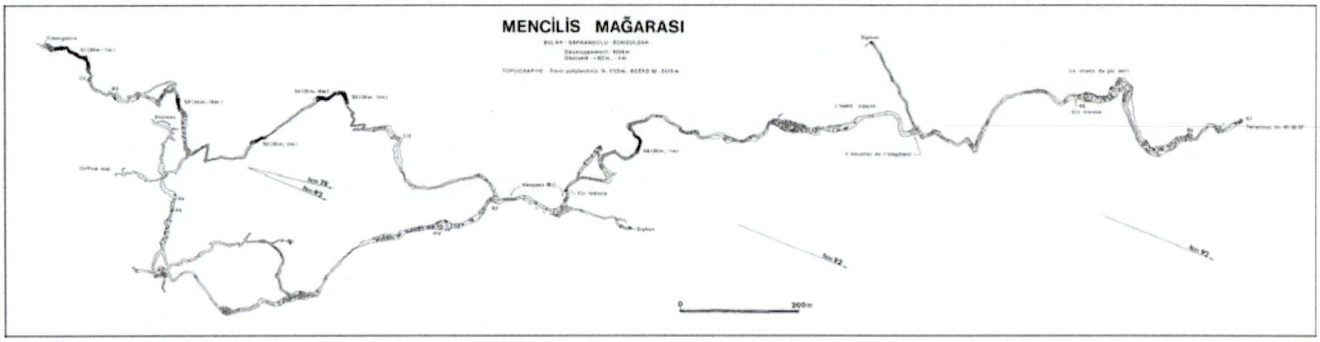

Fig. 4.9 Map of Bulak Mencilis Cave (Wolozan, 1992). Map of the first 2725 m of the cave is from (Watkins, 1980). Wolozan has an additional survey of 2525 m

Cave have continued till the seventh siphon. The first exploration of the cave started in 1977 by BÜMAK. It was explored and 2725 m of the cave was mapped by Trent Polytechnic in 1978. During this second exploration, bats belonging to Rhinolophus ferrumequinum, Rhinolophus hipposideros, and Hypsugo savii species were reported (Watkins, 1980). The fossil entrance and siphon exit were combined by ADEKS team in 1992. They also dived to the sixth siphon from the water inflow point and stopped at the seventh siphon. Furthermore, they carried out researches and tried to connect the four pits to Mencilis Cave which failed (Wolozan, 1992, 1993a, 1993c) (Fig. 4.9).

4.2.7 Cennet–Cehennem Caves

Silifke–İçel
Depths: −135 m and −110 m

Cennet and Cehennem Caves are 22 km away from Silifke. These two dolines, known since the ancient periods and referred to in numerous resources, have been formed as a result of a collapse in the Miocene calcareous formation commonly seen in the Mid-Taurus Range (Fig. 4.10). Two dolines are 75 m away from each other. Cennet, which is the larger one, has an altitude of 150 m from the sea level. It has an elliptical dimension of 275 m × 125 m with a total depth of 135 m. The doline is descended by a pathway with stairs remaining from the Roman Period. The remains of an ancient Byzantine Church are found on a big pile of remains and at a depth of approximately −75 m on the western end of Cennet doline. At this point, the cave starts and continues westward. At the deepest part of the cave, an underground river siphoning both ways can be seen and depending on the season, there is a chance of flood (Fig. 4.11). Most likely, Cennet has been formed as a result of the collapse of the roof due to erosion caused by this underground river. There are numerous different freshwater springs along the shore at this part of the Mediterranean and though a certain result has not been received from the tracer tests carried out until now, it is estimated as this underground river is flowing into the sea 7 km after sinking in the cave (Aygen, 1984).

Oldest reference about Cennet doline is from Strabo's Geographica, written around first century AD. In book 14,

Fig. 4.10 Aerial photo of Cennet (on the left) and Cehennem dolines (*Photo* THK)

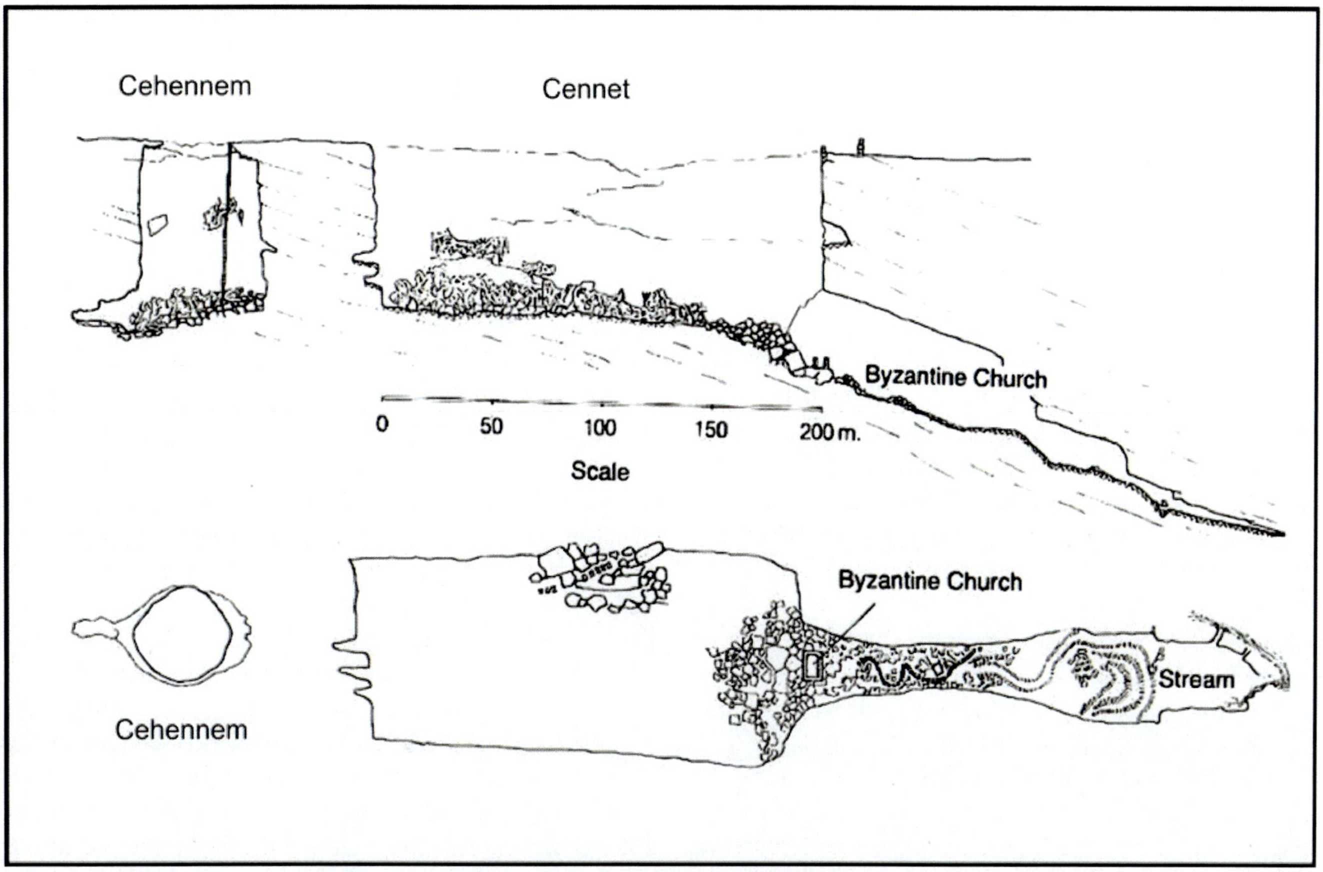

Fig. 4.11 Profile and plan of Cennet and Cehennem dolines, adapted from (C. Chabert, 1967)

Chap. 5, Sect. 5 of Geographica it was written that:... and to Crambusa, an island, and to Corycus, a promontory, above which, at a distance of twenty stadia, is the Corycian cave, in which the best crocus grows. It is a great circular hollow, with a rocky brow situated all around it that is everywhere quite high. Going down into it, one comes to a floor that is uneven and mostly rocky, but full of trees of the shrub kind, both the evergreen and those that are cultivated. And among these trees are dispersed also the plots of ground which produce the crocus. There is also a cave here, with a great spring, which sends forth a river of pure and transparent water; the river forthwith empties beneath the earth, and then, alter running invisible underground, issues forth into the sea. It is called Picrum Hydar. (Jones, 1929)

In this cave, the only change within 2000 years from the narrations of Strabo today is that there is no more a crocus in the cave and a church built during the Byzantine Period. It is also very surprising that Strabo, who has mentioned Cennet Cave in such a comprehensive way, did not even mention the other doline, Cehennem Cave at a distance only 75 m away from this doline (Yamaç, 2013a). Cehennem is an elliptical doline with a dimension of 50 m × 75 m and depth of −110 m formed as a result of a collapse. When compared with Cennet doline, it is steeper and narrower. The debris accumulated at the bottom of the doline due to the collapsed roof descends with a 30° inclination toward east from west.

4.2.8 Cumayanı Cave

Cumayanı–Zonguldak
Total Length: 1100 m
Total Depth: −16 m

Cumayanı Cave is located on Çatalağzı-Cumayanı road in Kilimli district of Zonguldak. The cave has been formed in the Lower Cretaceous aged Kapuz Formation which is the most karstic unit of the region. This formation, containing limestone with much fractured structure, is in touch with another formation formed by same-aged flysch just in front of the cave. The fault line passing just near the cave has pushed the limestone on the flysch and caused the occurrence of an impermeable layer. The numerous streams occurring as a result of the faults in the region have prominently cracked and fractured these Pliocene surfaces. And on the southern of Cumayanı Cave, there are karst areas and the most important of these karstic formations is Kabaklık uvala. The waters reaching this uvala enter the sinkhole from the western edge and after a certain distance, outflow from the fossil level of Cumayanı Cave.

The cave, as the last link of the system draining the surface and underground waters of a wide stream basin, has two separate entrances. There is the fossil entrance located below and 20 m above the mouth active with water outflow. These two separate entrances merge after a certain distance. The cave, with a total length of 1100 m, has developed in the direction of northwest and southeast from the start till the end of the big fossil hall. On the contrary, with a sharp turn after the dripstone bridge goes in the direction of northeast and southwest. Cumayanı Cave consists of two levels with different development periods. The fossil floor, the entrance of which is +20 m above the active water outflow and the Great Hall, as its continuation, is the first part of the cave (Nazik, Mengi, Özel, Bircan, & Beydeş, 1995). In the active gallery on the lower floor, there is a big underground stream and lakes in various sizes formed by this stream. Among the lakes, from place to place, islets of pebble and sand and huge blocks that have dropped from the ceiling are seen. The cave ends at a height of +16 m above the entrance with a very narrow siphon (Fig. 4.12).

The waters coming from Kızılelma Cave outflow from this siphon with high pressure. Though the ceiling height and width of Cumayanı Cave show significant differences from place to place, the width of the Great Hall on the fossil floor and Collapse Hall in the active gallery reaches 40–50 m. The surface and underground waters of the fluvio-karstic stream basin closed from the surface are drained with Kızılelma–Cumayanı cave system. Kızılelma is the starting part (sinkhole), and Cumayanı is the last part (spring) of the system. In addition to the basins, where the waters of Kızılelma Cave are collected, the water of Kabaklık uvala directly connects to Cumayanı Cave. The waters sinking into the sinkhole in this uvala outflows from the well located at the end of the Fossil Hall in Cumayanı Cave with high pressure. The waters outflowing from the entrance of Cumayanı Cave re-sinks in another sinkhole which is 10–15 m further below and after 450–500 m re-outflows in Cumayanı Plain (Nazik et al., 1995). Cumayanı Cave, a large part of which is located in the active/semi-active zone, hydrologically is poor in terms of dripstone formation. Apart from the wall dripstones in certain parts of the fossil gallery, the most characteristic formation of the cave is the dripstone bridge located on the connection point of the galleries. Dripstone pools in different sizes have been formed on the bridge consisting of white- and yellow-colored calcite crystals. The first speleological research of the Cumayanı Cave was carried out by Temuçin Aygen in 1975. During its second exploration by the Turkish and French cavers in 1977, the entrance and exit zones were mapped (Gilli, n.d.). The cave was entirely researched by the cavers from Trent Polytechnic University in 1978 (Watkins, 1980). The cave was researched and re-explored by MTA in 1995 (Nazik et al., 1995). During the studies carried out in 2011, seven different bat species have been reported.

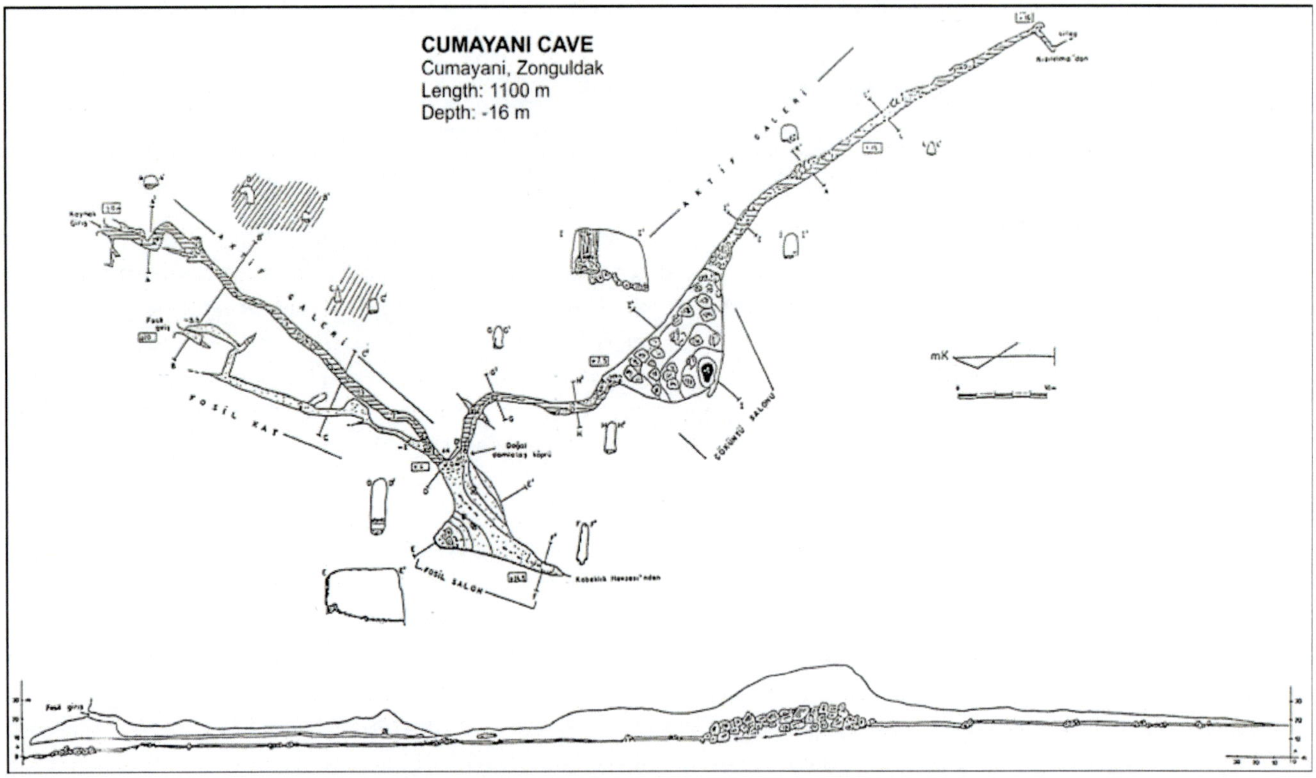

Fig. 4.12 Plan and longitudinal profile of Cumayanı Cave (Nazik et al., 1995)

4.2.9 Çem Sinkhole

Tomarza–Kayseri
Total Length: 1069 m
Total Depth: −605 m

It is neither the deepest sinkhole nor the deepest vertical pit of Turkey. For the time being, the deepest cave of Turkey is Peynirlikönü Sinkhole with a depth of −1429 m, and the deepest vertical pit is −305 m in Bucakalan Sinkhole with a total depth of −345 m. On the other hand, Çem Sinkhole is very different and interesting both in geological and speleogenesis terms. It is located 24 km east of Tomarza district of Kayseri in Çem Plateau in Arslantaş Village, is in the Mid-Taurus Range with an altitude of 2594 m. The cave has been formed in the Upper Miocene–Pliocene aged limestone outcropped on a wide area on the said part of the Taurus Mountains (MTA, 2002b) and is one of the caves with the highest altitude researched in Turkey. It is a known fact that during the cave explorations carried out on high altitudes on the Taurus Mountains, almost all sinkholes with wide entrances are reported to be blocked with snow (A Klimchouk, Nazik, Bayari, Törk, & Kasjan, 2016) and apart from Çukurpınar Sinkhole, all researchable deep sinkholes have a tight entrance. Moreover, in the Big Chamber arrived after the first big descent in Çukurpınar Sinkhole, there is an enormous ice accumulation. These sinkholes, draining the water of a huge basin on the surface through a fracture formed by the fault of the limestone, generally find a way from the fractures on the same fault. Though this physical erosion forms medium-sized chambers from place to place, these types of sinkholes mostly continue with narrow passages. Çem Sinkhole, starting with a small entrance disallowing any ice accumulation, continues till −52 m with small descents and is partially horizontal. Until reaching this depth, two small side branches connecting to the main gallery are most likely draining the groundwater from two different points. The enormous vertical shaft, with a diameter of approximately 45 m and depth of 277 m, encountered at −52 m is a very interesting geological formation (Fig. 4.13).

Though the shaft loses an amount of its width at −329 m and decreases to a diameter of 25 m, descents 109 m more and reaches −438 m. Afterward, it narrows a bit more and ends with a siphon at −605 m (Fig. 4.14). This shaft is not a fault mirror like the one in Keş Sinkhole and is most likely a formation formed with physical erosion. Çem Sinkhole does not have the catchment basin for such erosion and is a sinkhole on a mountain slope at a high altitude with a narrow entrance and horizontal start. The cave has been discovered, explored and surveyed as a result of the joint activity of three clubs; BÜMAK, BUMAD, and ITUMAK during 2010–2011 (Döker, 2013).

4.2.10 Çukurpınar Sinkhole

Anamur–İçel
Total Length: 3550 m
Total Depth: −1196 m

Çukurpınar Sinkhole is close to Olucak Village, 25 km north of Anamur district of İçel. The altitude of the sinkhole

Fig. 4.13 Beginning point of the main shaft of Çem Sinkhole (*Photo* B. Kurt)

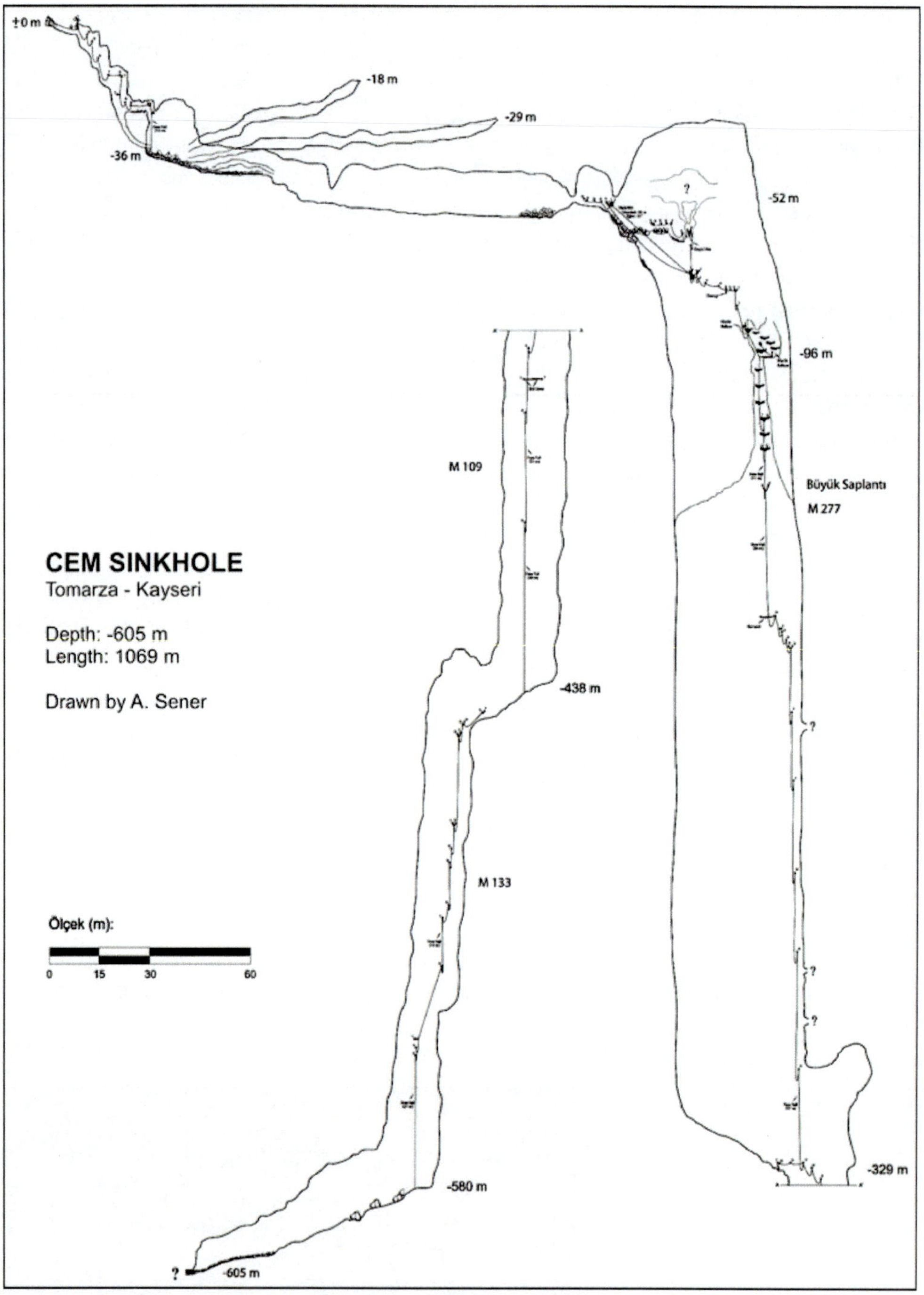

Fig. 4.14 Projected profile of Çem Sinkhole, drawn by A. Şener (Döker, 2013)

is 1850 m, and it is 500 m away from Peynirlikönü Sinkhole. Though it has the same geological formation as Peynirlikönü Sinkhole, Çukurpınar Sinkhole has a completely different structure. Concerning these sinkholes collecting an enormous amount of surface water, the tighter the entrance of Peynirlikönü Sinkhole is the wider the entrance of Çukurpınar Sinkhole is. Another interesting geomorphological characteristic is that the sinkhole of Çukurpınar, previously active, has been blocked with earth and mud. This previous entrance, which was the lowest point of the uvala, is in the form of a blocked doline. The surface water has eroded a new course in time and formed a huge entrance with a diameter of 30 m and a depth of 110 m (Usuloğlu & Siler, 2014) (Fig. 4.15). Below this first descent, there is an enormous chamber and a huge ice accumulation which is most likely never melting. The side branch of this chamber reaches below the blocked doline, which was the previous entrance and in this chamber, there are stalagmites formed by the water still leaking from time to time. The first 400 m of Çukurpınar Sinkhole is nearly vertical. This vertical formation slowly changes to horizontal after the first camp area at −515 m. Between −824 m and second camp area at

Fig. 4.15 First shaft of Çukurpınar Sinkhole (*Photo* M. Albukrek)

−1036 m (Aygen Camp), there are long galleries toward the southeast. The branch beginning at the second camp area is also horizontal. The cave starts to narrow and ends at −1196 m (Susam & Genc, 1996) (Fig. 4.16). Though this depth is roughly the base level of Taşeli Plateau and the starting level of the springs, the ending point of Çukurpınar Sinkhole is quite distant from any spring. Çukurpınar Sinkhole has been discovered by Temuçin Aygen in 1989, and its research has been completed by BÜMAK in 1994.

4.2.11 Damlataş Cave

Himmetli–Saimbeyli–Adana
Total Length: 317 m
Total Depth: −22 m

Damlataş Cave is located on the border of Himmetli Village, 18 km south of Saimbeyli. The area has an Ordovician–Upper Cretaceous aged autochthonous sequence, which yields a stratigraphy but very tectonized. The unit in which the Damlataş Cave is located was defined as the Dogger–Lower Cretaceous aged allochthone limestone. This unit is fine to medium-thick layered, generally dolomitized, less dolomitized, and less crystallized limestone. The study area became terrestrial due to the compressive tectonic regime during the Lower Carboniferous and became a re-deposition environment at the Upper Permian. The Early Cimmerian orogenic phase was dominated by regression tectonics at the end of Triassic. After a long period of erosion, at the end of Eocene, the region has gained a folded and fractured structure with a SE–NW directed tectonic forces (Tutkun, 1989). At the end of the Eocene period, the entrance of the cave, which was developed in the fractured–cracked limestone, was exposed due to the Göksu Fault.

Damlataş Cave has been developed along the north–south fracture line in the Upper Jurassic–Lower Cretaceous aged formations. It is 500 m higher (1100 m) from the karst basement level of the area (Göksu River, 600 m), and the entrance of the cave has a width of 10 m and a height of 2.5 m. About 10 m inside the entrance of the cave, there are travertine walls and columns. The next 40 m of the cave has also dripstones and columns, in addition to boulders of various sizes. The second chamber is the most ornamental part of the cave. This section is completely covered with columns, stalactites, and stalagmites. In this section, which is +2 m above the entrance of the cave, there is a chimney connected to the lower gallery in two steps of −3 and −10 m. The chimney is a connection between the upper and lower galleries. The final section has a ceiling height of 8 m and an average width of 4 m, and it is the highest part of the cave system. Since the section is completely covered with travertine deposits in this part, it is not possible to search further. The lower gallery, located at the bottom of the main gallery and reached by a chimney, has a depth of −22 m and also has the same N–S direction with the main gallery (Törk et al., 2007).

4.2.12 Dim Cave

Alanya–Antalya
Total Length: 360 m
Total Depth: −26 m

Dim Cave is 11 km away from Alanya on the Mediterranean Coast of Turkey. The altitude of the cave is 232 m from the sea level and is on the western slope of Cebireis Mountain with a height of 1691 m. Dim Cave has been formed and developed in Alanya Unit Formation formed by three overlapping nappes with different stratigraphy and metamorphism. Dim Cave, without any layer, has been formed on a quite apparent fault zone roughly in the

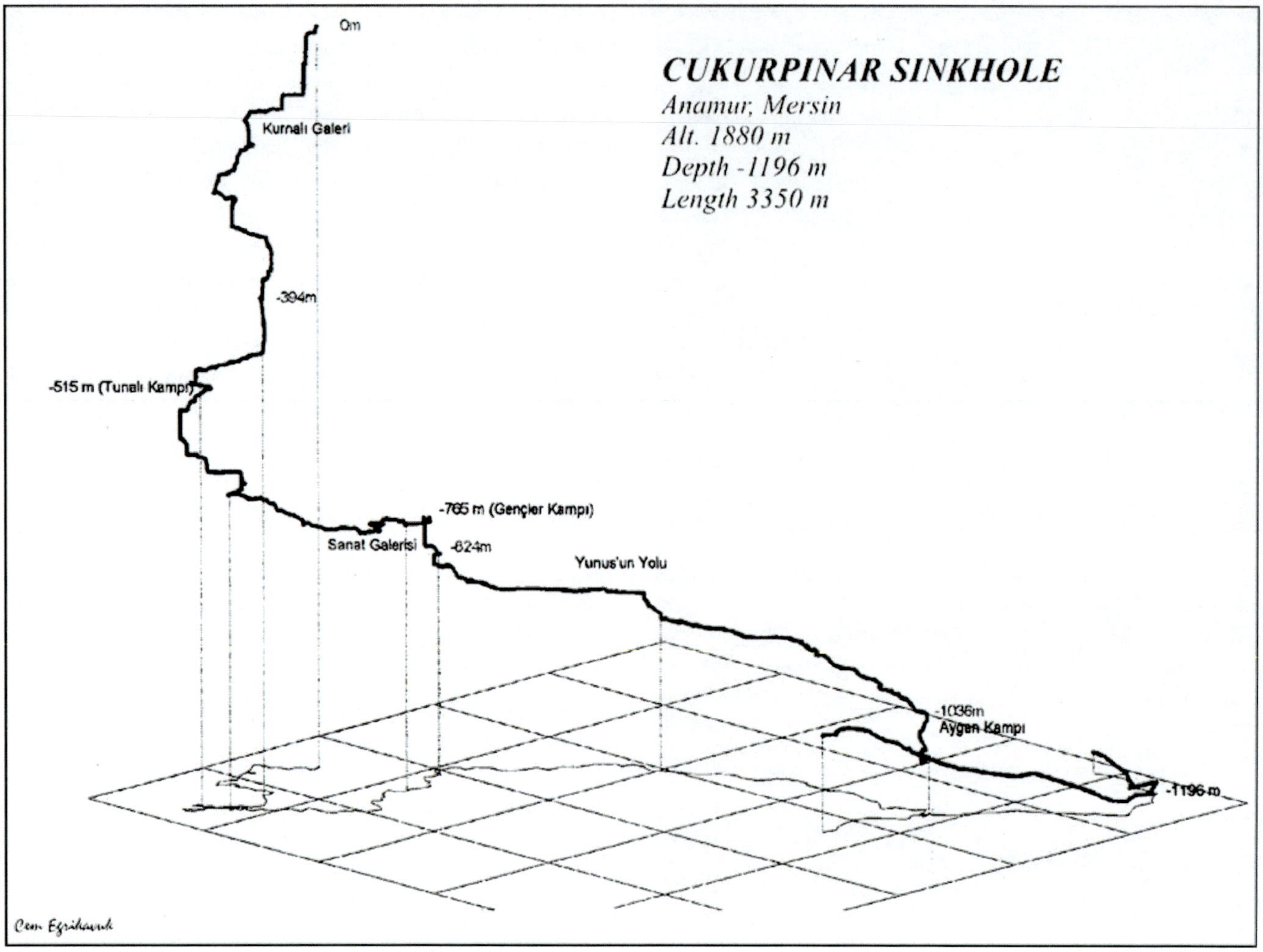

Fig. 4.16 Profile and plan of Çukurpınar Sinkhole (C. Eğrikavuk, 2006)

direction of north–south and from place to place in dolomitic crystallized limestone. Dim Cave has been formed in the early Paleozoic aged Central Taurids, a very solid and thin-layered dark gray-colored recrystallized limestone which composes the main structure of Cebireis Mountain (Güldalı et al., 1987).

The Taurids are on the upper Alpine orogenic belt and have reached their heights at the Pleistocene age due to the forceful tectonic movements. After these inclines, deep valleys and canyons have been formed as a result of the erosion caused by the streams. Dim River flowing in front of Dim Cave is similarly in a deep canyon. During the tectonic movements, Dim River deepened the riverbed to lower levels. Most likely, in the early periods, a huge underground creek flowing through Dim Cave was outflowing to Dim Valley. In the later periods, the water level lowered as well as a result of the gradual deepening of Dim River and Dim Cave dried up and turned into a vadose from phreatic. Though the cave had lost its hydrological activities, its volume expansion continued in the later periods with the collapse of the ceiling and rocks falling from the walls (Fig. 4.17). As a result of the water leaking from the ceiling and walls during the expansion, the cave is full of many kinds of dripstone formations (stalactites and stalagmites), and currently, the dripstone formation continues from place to place (Güldalı et al., 1987). Different arachnid species, worms, bats, and salamanders have been observed in the cave. Dim Cave has been opened to tourism in 1998 by a private company.

4.2.13 Dodurgalar Cave

Dodurgalar–Acıpayam–Denizli
Total Length: 145 m
Total Depth: −5 m +5.5 m

Dodurgalar Cave is located on the eastern slopes of the Mallı Mountain, 3 km west of the village of Dodurgalar.

The area around the Dodurgalar Cave is composed of different rock units, the lithostratigraphic, and structural features of the Lycian Nappes. The Cretaceous ophiolites

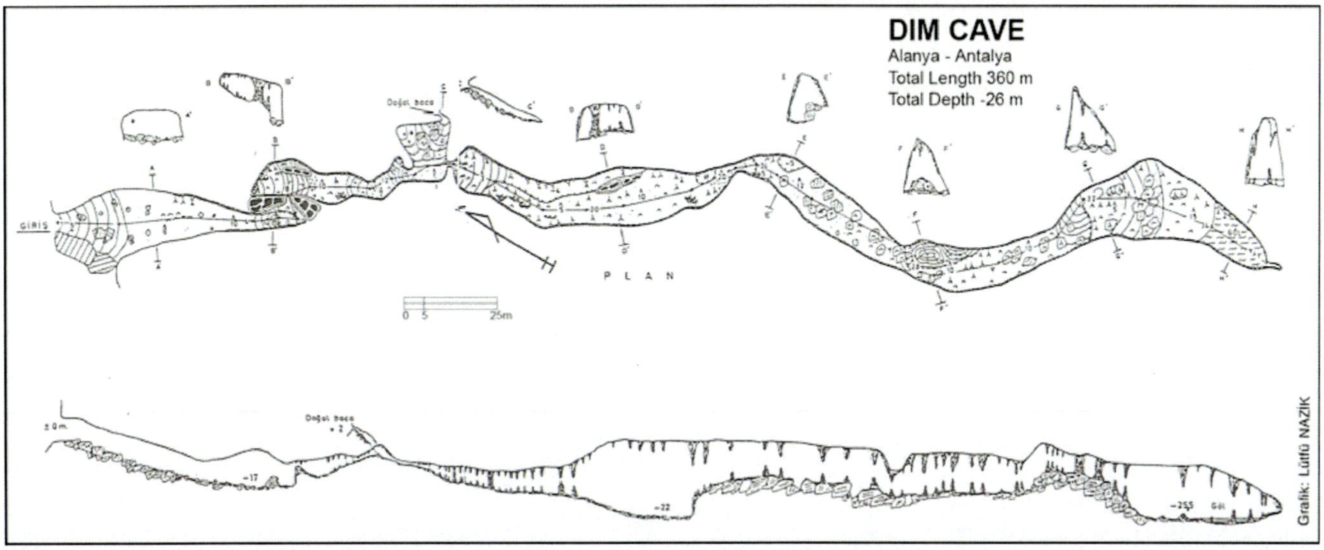

Fig. 4.17 Plan and profile of Dim Cave, adapted from (Güldalı, Nazik, Soylu, & Aksoy, 1987)

form the stratigraphic base from these rocks. On the ophiolites, limestone and blocky flysch were belonging to Liassic–Upper Cretaceous. The two rock units called Mallidağ Unit by Senel, 1977, and overlain by a tectonic contact. The Mallidağ Unit, the Middle–Upper Triassic flysch, and the Upper Cretaceous–Cretaceous calcite was overlain by the tectonic contact. Dodurgalar Cave formed in Jurassic–Cretaceous aged neritic limestone developed. The cave, which is approximately 200 m above the Acıpayam Polje (895 m), was developed in the north–south direction, just below the Pliocene surface.

The cave has a narrow entrance and a single chamber with a total length of 145 m. The cave, which developed in Jurassic–Cretaceous limestone, has a recessed protruding structure as it is divided into many rooms with many columns. An almost horizontal cave, the deepest point is −5 m and the highest point of the cave is +5.5 m. Dodurgalar Cave is in the vadose zone and was completely fossilized. Dripstone formations, and in particular, thick columns, prevent the danger of collapse. Water leaking from the ceiling during the rainy periods, as well as the new dripstones is kept alive the old formations (Fig. 4.18) (Nazik et al., 1998).

4.2.14 Döngel Cave

Döngel–Göksun–K. Maraş
Total Length: 319 m
Total Depth: +38 m

Döngel Cave is very near to Döngel Village, and it is an active aquifer. The cave was formed in Miocene aged reef limestone, and it is interconnected by two fracture systems within NE–SW and NW–SE directions. Döngel Cave, which was researched by the BUMAK, MTA, and OBRUK teams at different dates, has three different plans. This active cave, which is located at the bottom of the Döngel Cave System was explored by MTA in 2008, and it was measured as 117 m up to the siphon. During the OBRUK survey in 2010, a very long fossil branch that was not measured by the MTA was found and the cave was measured and mapped again. As a result of this study, it was understood that the total length of the cave was 319 m. The active part of the cave was developed within two intersecting joint systems. While the width of active passages varies between 2 and 3 m, the fossil branch is much wider and higher (Fig. 4.19) (Yamaç, 2011).

There is continuous water flow in the cave, and the average flow is around 2 L/s during the dry season. Water discharge must be much higher during the rainy season. The average air temperature is 12 °C, and relative humidity is % 92 in the cave during the summer. Even in the active part of the cave, few cave sediments are also observed. However, the hydrological structure of the cave prevents the formation of cave deposits (Törk & Savas, 2008).

4.2.15 Dupnisa Cave

Demirköy–Kırklareli
Total Length: 3150 m
Total Depth: +154 m

Dupnisa Cave is located 6 km southwest of Sarpdere Village of Demirköy district of Kırklareli. Though the Jurassic–Cretaceous limestone and marble outcropped on wide areas are observed on this part of Yıldız Mountains lined in the northeast of Thrace, from place to place, the Triassic schist underneath comes to light (MTA, 2002a).

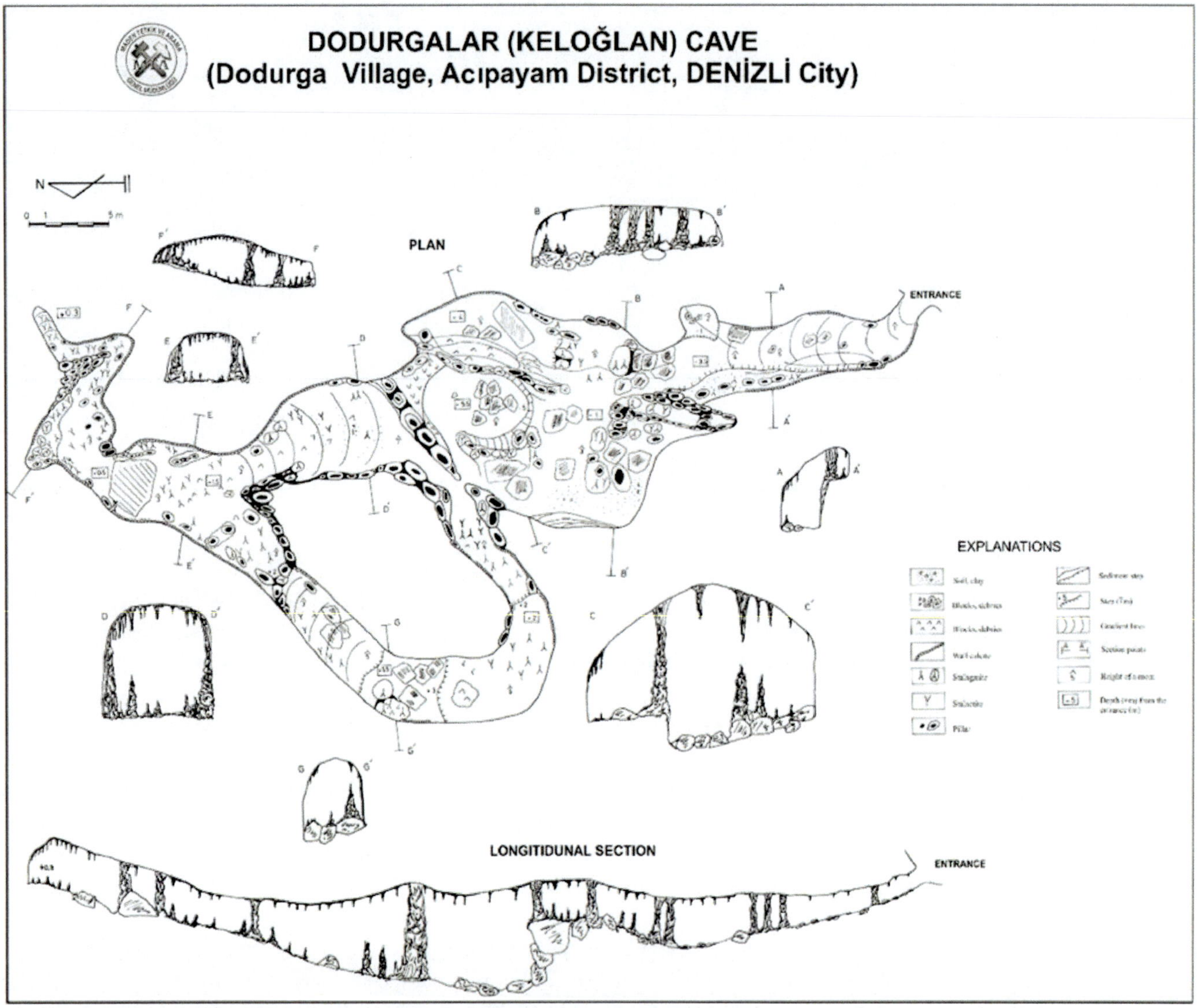

Fig. 4.18 Plan and profile of Dodurgalar Cave, adapted from (Nazik et al., 1998)

Dupnisa Cave is located on the border of the overlapping of the limestone with the schist and two different fault lines. Consisting of two vadose and one phreatic gallery, the cave system has a total length of 3150 m. The spring resurging out of Dupnisa Cave system, consisting of interconnected two levels and three caves, comprises Rezve Creek which is the border between Turkey and Bulgaria. The new entrance of Dupnisa Cave was reduced 20 m with respect to its former entrance due to a collapsed doline where the former entrance was replaced by a natural bridge. As progressed, the roof gets higher, and there are some large cavities. The water level is low in the cave (Fig. 4.20). From time to time, colored marbles covered with decorations can be seen. The fossil gallery right on top of the active gallery can be mostly followed. In the last part of the active gallery continuing for approximately 1 km, the dimensions of the large chamber are 125 m × 75 m, with a ceiling height of 30 m (Fig. 4.21).

Kızlar Cave is a fossil branch comprising the second level of Dupnisa system together with Kuru Cave. Very poor in dripstone decorations and consisting of three remarkable branches, Kızlar Cave has been completely fossilized. The bottom is a mixture of thick fossil soil and coarse blocks, pebbles, and sand, and it has two connections with Dupnisa Cave. Kızlar Cave starts with a 60° incline, and as progressed, it diverts into branches where the floors are covered with rocky blocks, and these branches leading to three diverse routes are then combined and joined to Dupnisa Cave. At this junction, there is 11 m descent from Kızlar Cave to Dupnisa Cave. Kuru Cave is the other branch of Dupnisa system together with Kızlar Cave. It has two entrances; the first entrance, 100 m to the southeast of Dupnisa Cave, is like a 5 m steep shaft while the second entrance is 225 m to the southeast and 12 m down of the first entrance. It is 456 m long, and the initial incline ends up with

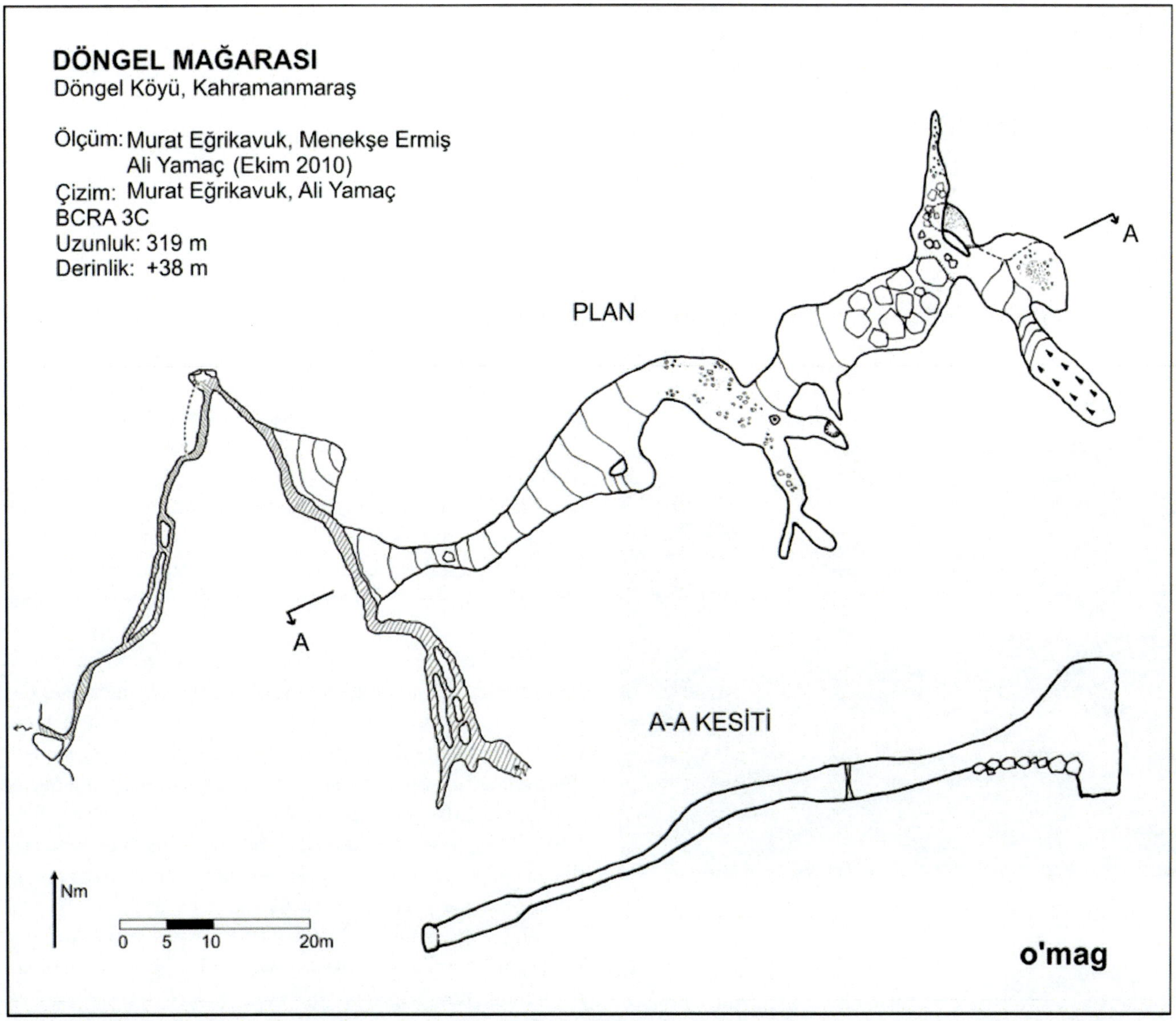

Fig. 4.19 Plan of Döngel Cave (Yamaç, 2011)

a room having a floor of soil and sand. One of the branches diverting from this room ends up with a chimney coming to the surface while the other one joins with Dupnisa Cave. The cave is adorned with decorations of stalactites, stalagmites, and wall dripstones. There are debris, blocks, and a thick soil deposit in the entrance zone. Small halls were developed due to depressions on parts covering Dupnisa Cave. Formed in the direction of north–south, a narrow gallery diverting from the eastern part of the Great Dripstone Hall descends to Dupnisa Cave. The speleological report of Dupnisa Cave System was first issued as a result of the explorations carried out by BÜMAK between 1979 and 1980 (E. Atalay & Ülkümen, 1980). In Dupnisa Cave System, a large bat population consisting of numerous different species lives are there. As a result of the researches and counting carried out in different years, it has been determined that the population differs periodically but regularly is between 30,000 and 50,000 (Benda & Horacek, 1998; Paksuz, Özkan, & Postawa, 2007; Zeinelabdin, 2002). Myotis myotis being the most common, 15 different species have been determined in the population. These are *Rhinolophus ferrumequinum*, *R. hipposideros*, *R. euryale*, *R. meheyli*, *Myotis myotis*, *M. blythii*, *M. emarginatus*, *M. mystacinus*, *M. capaccinii*, *M. daubentoni*, *M. bechsteinii*, *M. nattereri*, *Basbastella barbastellus*, *Miniopterus schreibersii*, and *Plecotus auritus* (Furman & Özgül, 2004). Dupnisa Cave system is the only known cave with 15 different species of the 35 total known bat species of Turkey existing together.

Fig. 4.20 Main gallery of Dupnisa Cave (*Photo* C. Güloğlu)

4.2.16 Gökgöl Cave

Erçek–Zonguldak
Total Length: 3350 m
Total Depth: −11 m +43 m

Gökgöl Cave is located on the road around Üzülmez Region at the 4th km of Zonguldak–Ankara highway, to the southeast of the city, in Ercek Village. Gökgöl Cave has been formed at a place close to the contact of two different formations composing Zonguldak Carboniferous and in the Lower Carboniferous limestone. The other formation, mainly composed of schist and sandstone alternation, is a hydrologically impermeable unit. On the contrary, the Lower Carboniferous limestone in this area has the lithostratigraphical characteristics suitable for karstification and cave development. The base of this limestone is marly and chalky, and as progressed to the upper levels, the dolomite ratio increases. On the top level, $CaCO_3$ is dominant. In the analysis of a sample, received from in front of the cave, 99% $CaCO_3$ has been determined. On the contrary, the amount of $MgCO_3$ is scarcely any. Concerning the formation of the cave, in addition to the lithostratigraphical characteristics, a fault in the direction of east–west has been effective as well. In most places, this fault is vertical to the dip of limestone. In some places, the cave extends in the direction of the fault and some other places in accordance with the dip of limestone. The developments in parts with asymmetric and narrow profiles are in the direction of the dip of limestone. On the contrary, the flat galleries with high ceilings have developed in the direction of the fault. Gökgöl Cave has three entrances, two of which have been fossilized and one is still active. After downpours, from time to time, water outflows from the smallmouth 10 m below the main entrance on the upper level. And the youngest entrance of the cave is located on the side of Ercek Creek. This third entrance is hydrologically active and very narrow. Apart from the main gallery in the direction of east–west, Gökgöl Cave consists of two big side branches. The main gallery, the endpoint of which is +10 m from the entrance, ends with a narrow siphon. The endpoint of the South Gallery diverging to the south from the Great Collapse Hall is +17 m, and the side North Gallery active diverging to the north with a wide arch

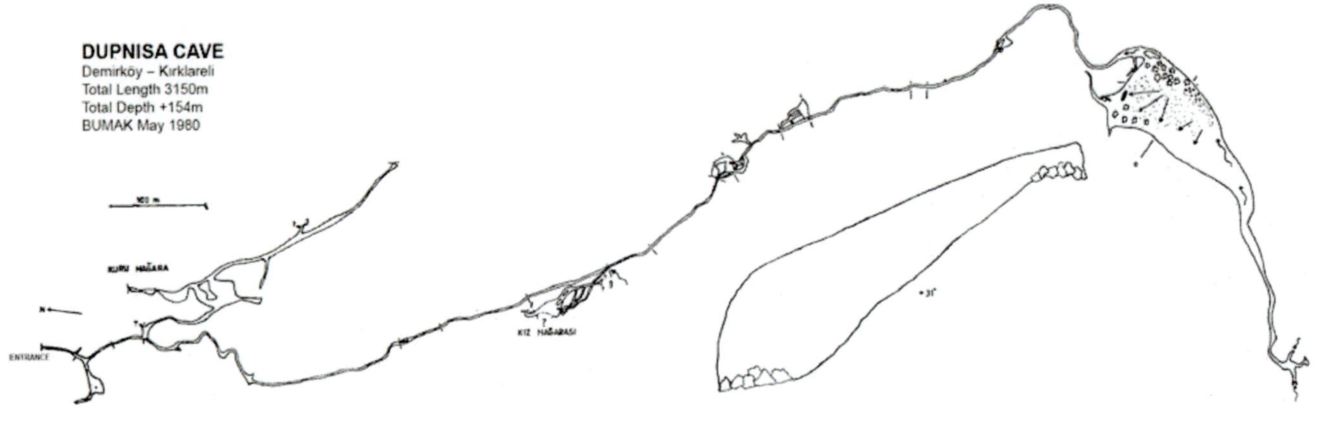

Fig. 4.21 Map of Dupnisa Cave (E. Atalay & Ülkümen, 1980)

is +21 m. The Great Collapse Hall is the junction point of all side branches. The width of the second part of the side branches and main gallery, located in the side zone hydrologically active, after the collapse is 1–5 m, and the height the ceiling differs between 1 and 8 m. On the contrary, the zone from the entrance to the Great Collapse Hall is wider and higher (Nazik et al., 1995) (Fig. 4.22).

Gökgöl Cave is located in four different geological and three different hydrological zones. The two fossil levels, characterizing the first development periods of the cave and located in the vadose zone, are entirely dry except for the waters leaking from the walls. On the contrary, the third and fourth levels are located in the active/semi-active zones. In the main gallery and side branches, even during the dry spell, there is an apparent underground creek. There is remarkable water inflow especially from the main branch diverging to the east. Though the levels of the water inflowing from the other branches quiet differ depending on the seasons, there is no significant change in the main gallery. Likewise, there is significant water inflow from the small branch active near the Great Collapse Hall. The waters inflowing from all branches of the cave merge in front of the Great Collapse Hall and disappear in the advanced active gallery due to a fracture in the semi-active main gallery. In the zone from the entrance to the siphon, there is significant water dripping from the ceiling and side walls during the rainy periods. The majority of these waters, forming ponds on the base of the cave, flow to the sinkholes at the entrance zone and outflow from the second mouth (Nazik et al., 1995).

Especially the fossil levels of Gökgöl Cave, rather rich in terms of the dripstone formations, located on the vadose zone, are decorated with dripstone formations as stalactites, stalagmites, columns, and curtains in all types and colors. On the contrary, on the active and semi-active levels, only wall dripstones and few stalactites are seen. The first 875 m of Gökgöl Cave, starting from the entrance zone to the Great Collapse Hall, has been opened to tourism. In this area, there is walking track, bridges, and observation platforms. The active underground creek continues to flow beneath the walking track. Its first exploration was carried out by the Turkish and French cavers in 1977. The cave was surveyed by cavers from Trent Polytechnic University in 1978 (Watkins, 1980). It was re-explored and re-surveyed by MTA in 1995 (Nazik et al., 1995).

4.2.17 İkigöz Cave

Çatalca–Istanbul
Total Length: 4816 m
Total Depth: +50 m

İkigöz Cave is located near Pınarca Village approximately 50 km north of Çatalca district of Istanbul. The outflow of the cave lies 150–200 m northwest of Pınarca

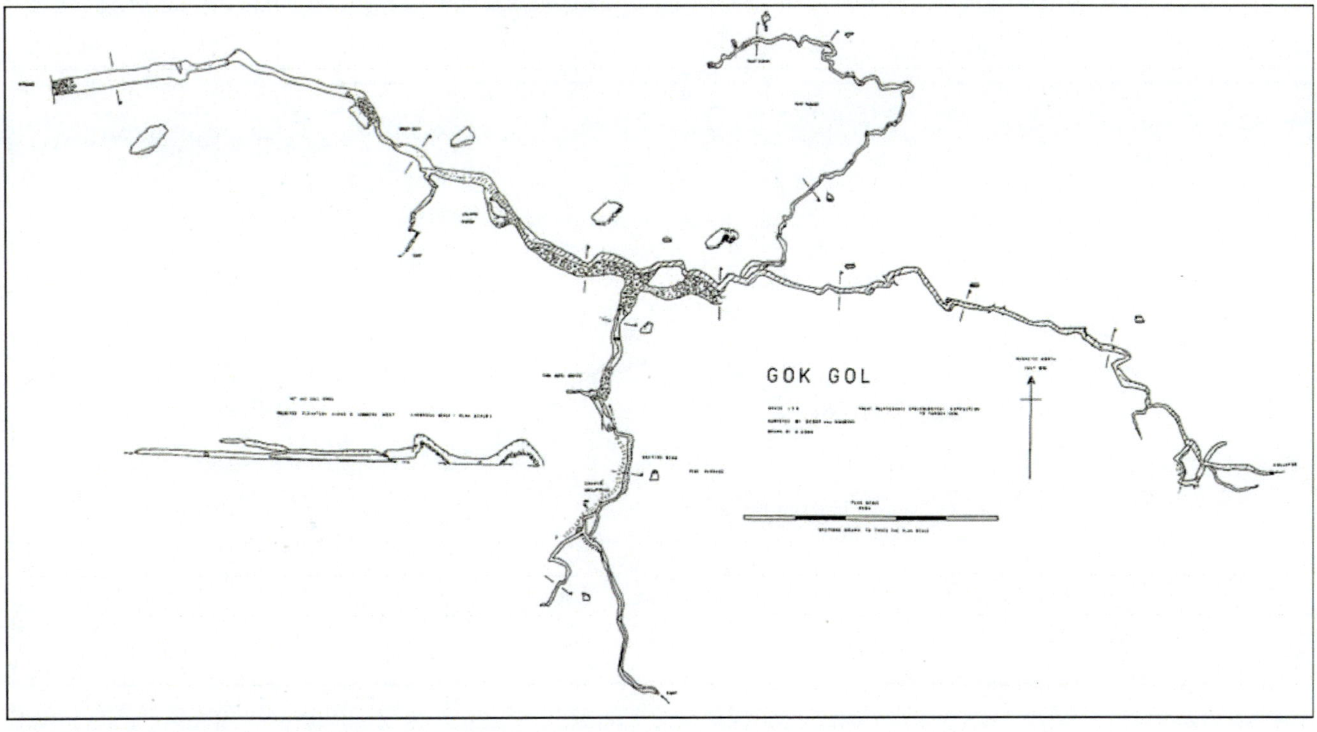

Fig. 4.22 Plan of Gökgöl Cave adapted from (Watkins, 1980)

Village. It can be reached by following the creek flowing through the village. On the way to the cave, the ancient Byzantine water tunnels can be seen. The unbroken parts of these tunnels lead to the cave after 35 m. As the first part from the entrance up to the chimney has two mouths, the cave is called "İkigöz" (Two Eyes) by the local people. The continuation of the cave after the chimney is hardly noticed.

This part of the mid-region of Thrace is rather lithostratigraphically complicated. In the region, there are at least 11 different geological formations and around İkigöz Cave, and two different formations are intersecting. These Middle and Upper Eocene aged rocks formed the cave, named Soğucak Formation, and another Upper Eocene and Early Oligocene aged formation are transitive. The rocks are mostly formed of carbonate shelves. The limestone is medium to thick-bedded and massive, has high porosity, melting gaps, and fossiliferous. This formation with a very slight thickness and the other transitive formation containing pelagic shale, marl, and clay have the characteristics of the impermeable layer beneath the limestone (Altiner, Bati, Tunay, Senel, & Ekmekçi, 2006).

The accessible part of İkigöz Cave is the active gallery with a length of 130 m and a height of 1.5–2 m. A stream flows through the gallery. Except for a few dunes, the floor of the cave is 0.5–1 m below the water level. At the end of the gallery, a collapsed chimney forms another entrance to the cave. The underground system passing through the cave system is the main drainage system of the area, and it had been used as an important water source in the past. There are remains of the ancient dike connected to the cave, 2.5 m above the chimney. After passing the chimney, it is quite hard to proceed further in the cave. The water beneath the low entrance forms a siphon in the winter months. In summer, the water level drops and enables a low passage. From this point after, the cave continues for another 100 m and reaches an inclined chamber with a permanent siphon at its end (Özgül & Bilgin, 2000).

During the two explorations in 1990 and 1992, the cave system hidden behind the siphon was explored by the cave divers. The first and second siphons were dived during the explorations. Three hundred m after the second siphon, the gallery enlarges. The width of this section, namely "Gallery of Atatürk" ranges from 8 to 15 m and its height ranges from 10 to 20 m. Afterward, the gallery is divided into two. In the area, namely "Du Cote d'Ailleurs," the ceiling is full of thin soda straws in an area of 400 m until the third siphon. In 1993, the third siphon was dived and the total length of the cave reached 4816 m. Since the endpoint of this gallery is quite close to Kocakuyu Cave with a length of 1010 m, it has been indicated in a report as Kocakuyu Cave might be a former branch of İkigöz and both caves might be a part of the same cave system (Fig. 4.23). As some of the siphons remain unexplored, there is a possibility that these two caves might be directly connected (Wolozan, 1993a, 1993b, 1993c).

As some bats have been seen in the galleries between the siphons, there may be some hidden cracks or chimneys opening outside but undiscovered yet. Due to their sizes and isolations, the galleries might provide excellent roosts for and host large colonies of bats. The tooth of a mammoth semi-covered by the flooded wastes was uncovered in the Gallery of Atatürk. And, two small fish were seen in the first siphon. Plenty of cave pearls were found in some chambers (Wolozan, 1993c). It was

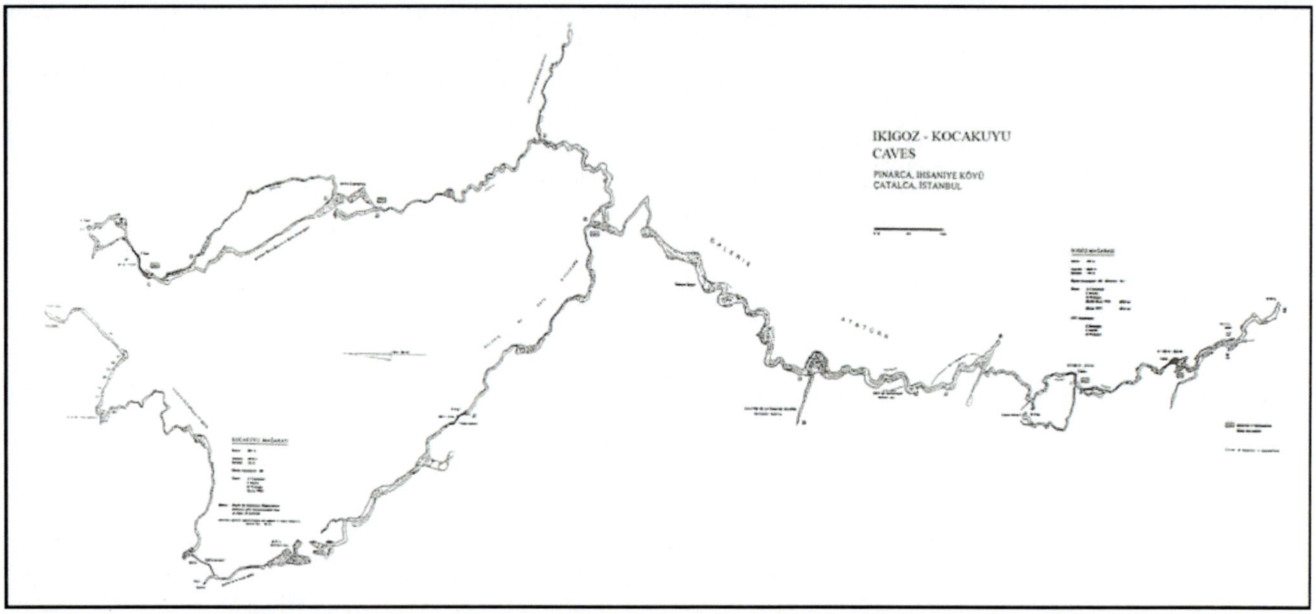

Fig. 4.23 Maps of İkigöz and Kocakuyu caves. At the bottom left corner of the map, Kocakuyu Cave can be seen, and the estimated distance between these two caves is less than 20 m (Wolozan, 1993c)

explored biospeleologically in 1999 within the framework of the Eurasian Land-Bridge '99 Project, and a few Rhinolophus Euryale was seen (Özgül & Bilgin, 2000).

4.2.18 Ilgarini Cave

Sümenler–Kastamonu
Total Length: 858 m
Total Depth: −250 m

Ilgarini Cave is on the north of Sümenler Village of Pınarbaşı district of Kastamonu. To reach this cave located in Küre Mountains National Park, approximately 4 h of walking through the forest is required, and since the forest is quite dense, it is hard to find the cave without a guide. The starting point of the forest, at the same time, is the intersection line of Ulus Formation constituting this area and Inaltı Formation rising as a hard slope on the north. From these two formations, both of which are Upper Jurassic–Lower Cretaceous dated, and Ulus Formation has created a sort of subduction zone from south to north by going beneath Inalti Formation. This subduction zone can be seen from the entire south wall of Küre Mountains National Park. This zone forms the boundaries of the national park. Due to the sinking of Ulus Formation beneath Inalti Formation, it has raised the other formation and formed a plateau entirely formed of chalkstone.

Ilgarini Cave, which is located on this plateau, was explored by Prof. Cemal Arif Alagöz in 1940 while carrying out researches in the area. In his work published in 1944, Alagöz has stated that they reached Ilgarini Cave with great difficulty and described the chapel and tombs in the cave (Alagöz, 1944). Ilgarini Cave was surveyed and mapped by BÜMAK in 1982 (Çetin et al., 1983) (Fig. 4.24) and afterward in 1990 by Leicester University Caving Club (Holland,

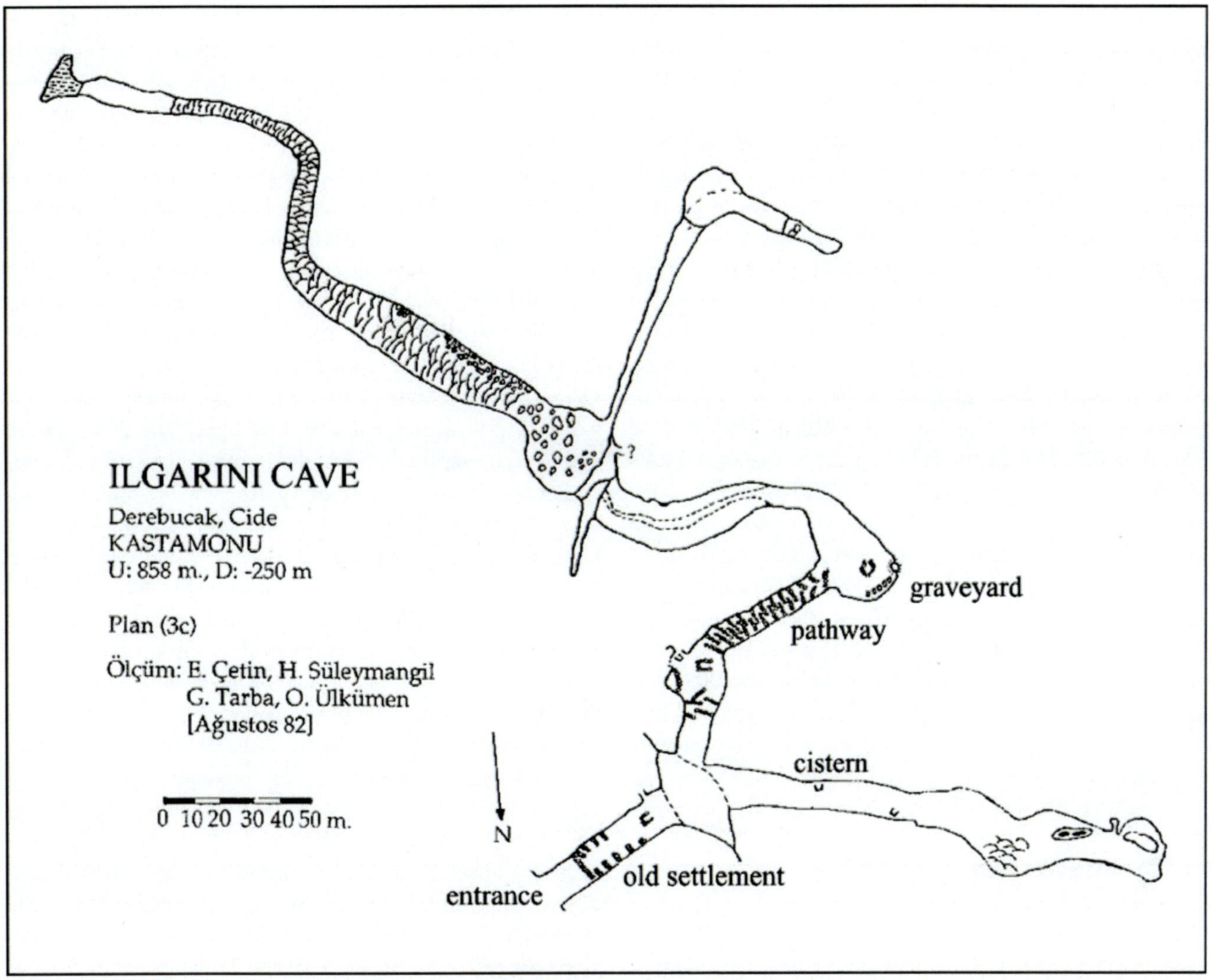

Fig. 4.24 Plan of Ilgarini Cave (Çetin, Süleymangil, Tarba, & Ulkümen, 1983)

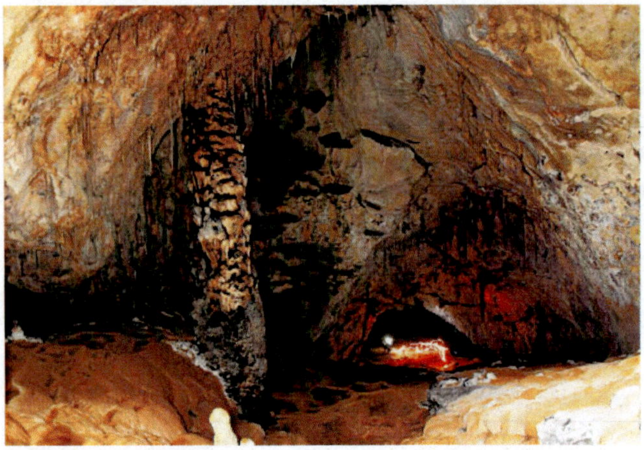

Fig. 4.25 Entrance gallery of Ilgarini Cave (*Photo* A. E. Keskin)

1990; Kay, 1990). The cave consists of two parts as fossil horizontal part and an active vertical part. In the horizontal part with a length of approximately 400 m, there are the remains of graves, houses, and cisterns (Fig. 4.25). The flat area just to the west of the cave entrance is supported from the west side by a retaining wall made of stones. Therefore, it can be concluded that this flat area is man-made. This flat area might have been created for agricultural purposes. During the studies carried out in 1982, on the enormous entrance of the cave receiving daylight, a ruined village of homesteads was reported (Çetin et al., 1983). As progressed to the left from the shaft to down below, following a zigzagging brick laid pathway supported by the stone setting walls, there is another chamber with two small churches and 11 graves. Most of the tombs are two–three layered with wooden planks used to separate. As a result of the observations done in September 2000 and May 2001, it is determined that these tombs have been greatly disturbed since the research of BÜMAK in 1982. The dendrochronological tests on the wooden samples taken for determination have revealed that the woods used on the tombs were mostly from oak trees common around the cave and approximately dated AD 977 (Akkemik, Aytuğ, & Güzel, 2004). Based on these, it has been understood that the tombs and other archaeological remains belonged to the late Byzantine period. On the other hand, a biospeleological research carried out in 2009, it is found out that the cave is home to a huge variety of biological life with endemic species from Pseudoscorpionida and Amphipoda group (Yamaç & Eğrikavuk, 2010).

4.2.19 İnsuyu Cave

Burdur
Total Length: 8350 m
Total Depth: +156 m

Insuyu Cave has located 13 km from the city center to the east of Çine Plain, southeast of Burdur, on the outskirts of Sarpgüney Hill with a height of 1606 m and is the first show cave in Turkey. This area on which the cave is located has a quite complicated geological structure. The lithologic units observed near the cave can be examined under two main titles as autochthonous and allochthonous units. The autochthonous units principally consist of Upper Cretaceous aged Söbüdağ and Senirce Formations, Plio-Quaternary aged travertine and alluvium sediments. The structural anomalies, which are the most basic characteristic of the high tectonic activation in the area, are frequently seen in the lithostratigraphical sequence. And the most determinant is Gökçebağ Ophiolitic Melange, the representative of the Lycian Nappes, constituting the allochthonous source of the area. The stratigraphical positions of the units to each other are determined by Fethiye–Burdur fault. Söbüdağ Formation is the oldest autochthonous unit near Insuyu Cave. The formation, also named as Söbüdağ limestone in the previous studies, is outcropped on a wide area. The formation has a very faulted structure and has a thickness of more than 500 m. The age of the formation is stated as Upper Cretaceous. The same-aged Senirce Formation, just above Söbüdağ Formation, principally consists of pelagic limestone. In the base areas with relatively high carbonate content, the formation shows bedding with a slight or medium thickness. These two autochthonous units show a structural unconformity in terms of the allochthonous Gökçebağ Ophiolitic Melange. The settlement age of Gökçebağ Ophiolitic Melange, the representative of the Lycian Nappes, is stated as Upper Cretaceous–Lower Paleocene (Karaman, 1994). The rocks composing the complex are principally serpentinite, gabbro, diabase, chert, radiolarite, and sandstone and limestone blocks in different sizes and ages. The unit with an extremely complex internal structure overlaps to the autochthonous Söbüdağ and Senirce formations around Insuyu Cave (Erdogan, Ergeneli, & Berberoglu, 2014). And these units, as of the entrance of Insuyu Cave, are covered with Plio-Quaternary sediments in the direction of the southwest–west–northwest. These sediments are alluvium, alluvial cones and fans, and travertine with a thickness of 300 m from place to place. The formation of Insuyu Cave is multiphase. The distribution of dry galleries to the subbranches and erosion formations which can be observed on the ceilings and walls of these parts only possible to be caused by high water flow indicates that these galleries currently dry have developed under submerged conditions. The cave probably transformed from saturated to semi-saturated conditions with time. During this process, the subterranean streams and branches with high flow speed and turbulence were also effective in the erosion. Probably, Insuyu Cave used to discharge much more amount of water than observed now (Fig. 4.26). Taking into consideration the morphology

Fig. 4.26 Diver in the first lake of İnsuyu Cave (*Photo* A. E. Keskin)

of the "keyhole" in some narrowing parts of the tributaries, semi-saturated following the primary waterlogged development period, and stream erosions are other pieces of evidence of the secondary development phase. The spoon formations on the high parts of the walls and ceilings in these parts show the rise of the stream from time to time filling the galleries with water. The presence of formations such as soda straws and stalactites are sufficient evidence to prove that the erosion by the subsurface stream reduced and lost its effect in time.

These deposits are the products of the latest development phase which has been effective up till now. The falling of the blocks from the ceiling commonly encountered in most parts of the cave shows that the tectonic movements are not only effective in the development of the cave but also in the destruction as a result of the natural process. The parent rock forming the walls in some parts is extremely diffusible because the corresponding blocks are intensively affected by the shear through the fault scarp notable in the development (Bayarı, Varinlioglu, Keskin, & Erdem, 2002).

Upon the informing of Dr. Temuçin Aygen, the scientific exploration of Insuyu Cave began in 1952. Having realized the importance of the cave after his first exploration, Aygen draws the attention of the authorities and carried out a second and comprehensive exploration in Insuyu Cave in 1953. Aygen mentioned in his book "Türkiye Mağaraları" (Caves of Turkey) that before the opening of the artificial tunnel, which provides entrance for tourists nowadays, it was quite hard to enter the cave from the natural entrance. Besides, the natural opening was completely submerged during spring with the increase of the water level (Aygen, 1984). Insuyu Cave was opened as the first touristic cave of Turkey in 1966. A biospeleologic research by Paolo Marcello Brignoli (Brignoli, 1971, 1972) and research for bat species by Friederike Spitzenberger were also carried out in the cave. The several bat species inhabiting the cave were determined as *Rhinolophus euryale*, *Rhinolophus meheyli*, *Rhinolophus blasii*, *Myotis myotis*, *Myotis capaccinii* and *Miniopterus schreibersii* (Benda & Horacek, 1998). Also, 200–300 hibernating Myotis capaccinii and 50 Miniopterus schreibersii were found (Spitzenberger, 1973). Also, Millipides and Troglobitic isopods in addition to a few species of spiders were reported (Brignoli, 1973).

The first comprehensive map of Insuyu Cave was drawn in 1968. Only the main galleries and some major branches were shown on this map, and none of the small galleries were shown. However, the narrow passage, providing a pass toward the non-touristic part of the cave and located on the northeast of the Big Lake, was shown on this map (Nazik, Derici, & Kutluay, 1999). This passage, which had not been researched for long years, was first passed in 1993. In the report by the research team, being the first to explore this part, the cave was stated to continue with several galleries from that part on and to have many lakes. A comprehensive map of this non-touristic part was completed in 2006. On the other hand, the modification of the map for the touristic part prepared in 1997 and re-drawn maps of the same part were also rather insufficient. OBRUK Cave Research Group realized the deficiencies in those five maps and started a project in 2011 to provide an accurate mapping of Insuyu Cave as much as possible. During this project, seven survey trips were made and a new map of the cave with a length of 8350 m was drawn (Fig. 4.27). The relevant length is almost double of all previous surveys. During the explorations, all galleries in the touristic parts of Insuyu Cave have been re-surveyed. Additionally, in the second part of the cave, a new water gallery and another large chamber following that gallery have been explored (Yamaç, 2013d; Yamaç & Eğrikavuk, 2013).

4.2.20 Kadıini Cave

Alanya–Antalya

Located on the Gömeme Mountain in the Oba Village, this cave lies 15 km northeast of Alanya. The cave is on the western rocky wall of Oba Çayı valley and 40 m above the brook. Oba Çayı deepened its bed by constantly eroding the Upper Permian aged dolomitic limestones that compose the area also known as the Cebireis Formation, causing the caves formed by karstic aquifers feeding this creek to stay in a hanged state (Siler, 2016). The underground water flowing in the Kadıini Cave also deepened its bed over time. Today this underground water continues to flow approximately 20–30 m below the cave entrance, goes through a short siphon, resurfaces through Kadıpınar Springs 120 m east of the cave, and joins the Oba Brook. The hole formed by the erosion of this underground water and the collapse of the weakened ceiling of the cave created a large chamber at the

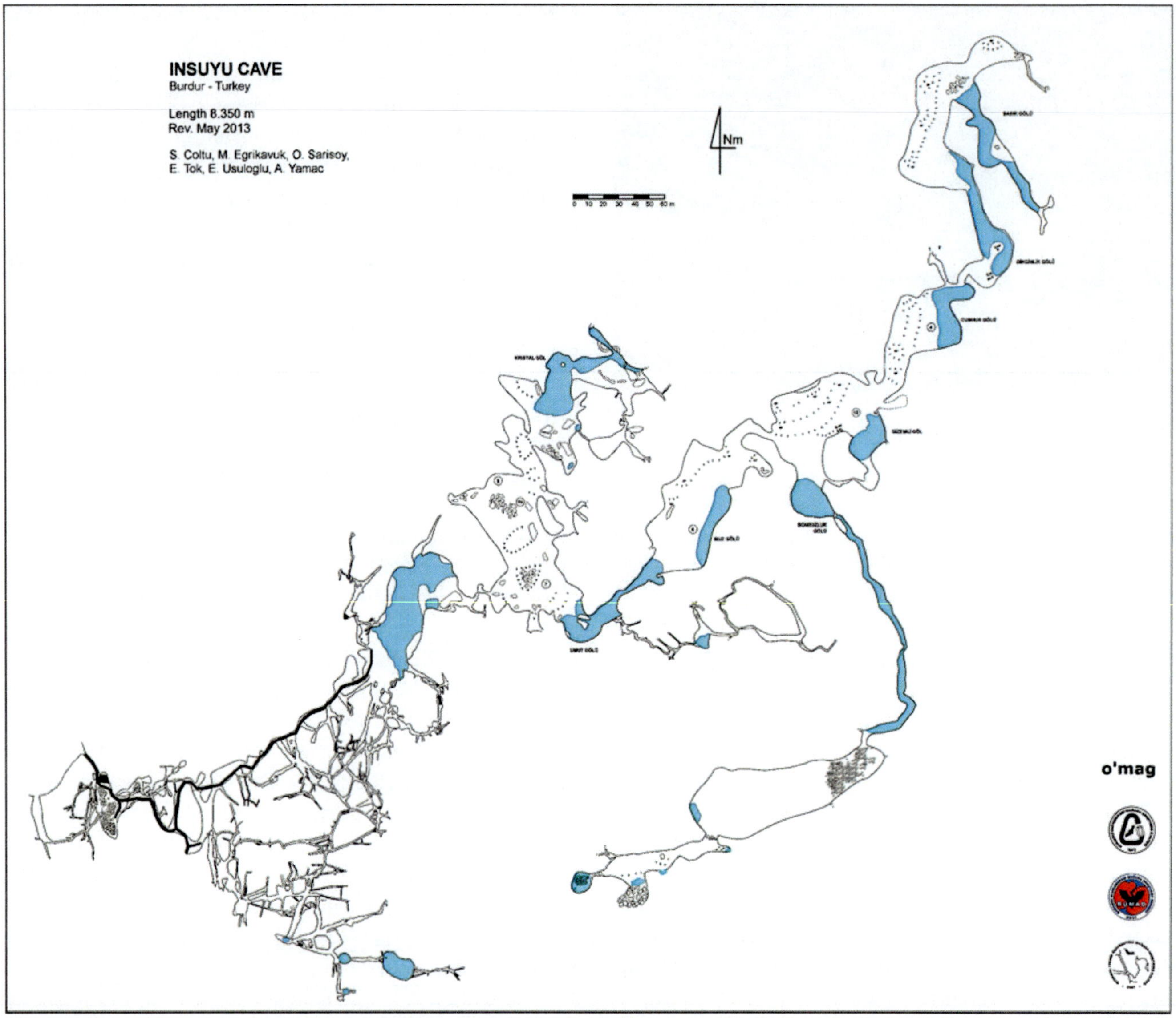

Fig. 4.27 Map of İnsuyu Cave (Yamaç & Eğrikavuk, 2013)

entrance of the Kadıini cave. The base of this chamber, approximately 120 × 40 m in size, is completely covered with big boulders fallen from the ceiling of the cave (Güldalı et al., 1987). There is one hard to find and tight side gallery between the boulders at the north wall of the chamber, which continues for more than 2000 m enwidening. Tens of human skeletons and pottery were found in a small chamber 120 m further from the starting point of this gallery almost completely covered with stalactites, stalagmites and travertine formations. Probably due to the difficulty of finding the entrance of that side gallery, these remains were undisturbed until found by the cavers of AKUMAK (Akdeniz University Caving Club) very recently.

Kadıini Cave was archaeologically studied for the first time by Kılıç Kökten in 1956–1957. Kökten opened up a trench in the main chamber close to the entrance, and in this test dig, he revealed that Classic Age, First Bronze Age, Neolithic Age, and Upper Paleolithic Age layers were found. He also found human bones in the same trench (Kökten, 1958). Even if it is known that the cave is inhabited for a very long period, in light of these excavation findings and some other fragments found in the main chamber of the Kadıini Cave, skeletons and pottery found in situ in this side gallery are important archaeological discoveries. Being cemented by calcium carbonate on the calcareous blocks, some skulls easily dissolve when touched, whereas some are silicified. Pottery and skeletons found by archaeologists were gathered by Alanya Museum in 2017 (Fig. 4.28). Kadıini skeleton group, which is a rare example of using caves as a burial ground tradition, is dated back to Late Chalcolithic and Early Bronze Ages according to the archaeological findings found along with them, which

Fig. 4.28 Bronze Age pottery fragments in Kadıini Cave (*Photo* E. Gilli)

coincides with the datings revealed by radio-carbon tests conducted on skeleton remains found in the previous studies. Large and small pithoi and urns, a tankard, a whorl, few needles, and stone tool remains were found inside the cave along with the aforementioned skeletons. It is speculated that pithoi and urns were used to bury babies and children. Anthropological and archaeological data shows us that these parts of the cave were probably areas used for the dead to be left and/or buried. It is speculated that the dead are thought to be left directly inside the cave or buried very close to the surface. On the other hand, most of the bones were recovered from wide pits, which probably indicates that the dead were left on berms first and moved to these pits later to make room for new ones (Usta & Ipekoglu, 2019).

4.2.21 Karain Cave

Döşemealtı–Antalya
Total Length: 100 m
Total Depth: −15 m

Karain Cave is approximately 27 km northwest of the city of Antalya, immediately northeast of Yağca Village. This small cave contains some of the most important archaeological findings of Turkey. The cave, which lies in the west Taurus Mountains system, within the Bey Mountain's Cretaceous aged limestones. The cave is 430 m high from the sea level, 110 m high from the plains in front of it, and opens up with a faultline. There are many karstic springs like the Kırkgöz springs and many different archaeological caves at the skirts of the hillside where Karain Cave resides. The cave, where the first human traces are dated back to the Pleistocene period, appears to be in a suitable condition in terms of ecological environments and inhabited for tens of thousands of years. The cave consists of three different spacious galleries and the five small sections at the first big space.

The Karain cave site was discovered by Moretti (1923–1924a). Even if he mentions it as the "Sacred Cave of Iuvadja" in his article, we know that this cave is Karain. In his study, Moretti was content with copying just the inscriptions inside the cave. The next study in Karain was conducted by Kılıç Kökten in 1946 as a side survey, during his research at the Gurma (Kurma) Cave. Kökten has decided to excavate the cave as a result of the chipped stone finds he encountered on the present floor of the cave. The excavation was started in 1946 in the most illuminated Chamber A (Kökten, 1955). The excavations continued until 1972 under Kökten's direction. In the excavations, he conducted in an area of 6 × 7 m, Kökten determined 8eight cultural layers and dated these layers to various periods from the first stages of the Lower Paleolithic Age to the depths of the Upper Paleolithic Age. Following Kökten's death in 1974, excavations were halted for over a decade. The trenches were reopened in 1985 by Işın Yalçınkaya; who set out to solve some of the problems in Kökten's stratigraphic sequencing; to further understand the chipped stone sequences; the Pleistocene faunal and floral assemblages of the site (Yalçınkaya, 1987). She also analyzed the sedimentary layers for age estimates. Today, the studies continue with the contribution of international experts.

Starting from the Lower Paleolithic Age, the cave had been consistently inhabited during the Middle and Upper Paleolithic, Neolithic, Chalcolithic, Early Bronze Ages, second century BC, and Classic Age. The inscriptions by the entrance indicate that the cave was used as a cult center in the Hellenistic and Roman periods. A cultural deposit of about 11 m thick was excavated, and it was understood that the longest and most significant settlement was during the Paleolithic Age.

The excavations directed by Yalçınkaya were carried out in the same outermost chamber that Kökten excavated in; Chamber E, as well as in Chamber B immediately to its west (Fig. 4.29). The excavations in the latter chamber exposed 12 geological strata and 32 archaeological layers. Of these 32, Paleolithic begins after the thirteenth layer. Layers 15–17 are Late Upper Paleolithic while layers 18–32 are Middle Paleolithic (Yalçınkaya, 1987). A sample from the Late Upper Paleolithic was dated to 16,250 ± 790 BP. Excavations in Chamber E included a vertical strip going along the badly damaged eastern profile and a horizontal one going along the remaining depositional layer. The lowest and oldest phase yielded some roughly denticulated and notched tools which are believed to be Clactonien in character. This phase has been dated to pre-350,000 BP. The phase above is comprised of units IV and III and is dated between 350,000 and 300,000 BP. This layer yielded tools including side scrapers with stepped and scaled edge retouch which

Fig. 4.29 Excavation in the entrance gallery of Karain Cave (*Photo* B. Erdem)

resembles European Charentien types. Roughly denticulated and notched tools continue to be found in this phase. The uppermost phase, comprised of units III, II, and I, have been dated between 300,000 and 60,000 BP (Yalçınkaya, Otte, Taşkıran, Kösem, & Ceylan, 1997).

There are countless flints among the Middle Paleolithic findings found within Chamber E. Disks and discoid cores as well as end scrapers are tools that collectively represent this period (Yalçınkaya, 1992a). The excavation yielded high numbers of flakes and debitage products, indicating that flake removal was carried out within the cave. Smaller flint cores seem to have been preferred in this period, instead of larger ones. The fact that they reused Levallois cores as discoid cores shows that they were economic in the use of this raw material. The points and end scrapers show that the inhabitants of the cave were hunters. A tooth mark, a mandible fragment, phalanxes, and an in situ leg bone belonging to a Neanderthal were among the findings. It is not clear whether all of these bones belong to one single Neanderthal or not (Yalçınkaya et al., 1998). All of these finds encountered in Chamber "E" belong to different phases of the Middle Paleolithic. It is suggested that Chamber B was used for butchering activities during this level, based on the animal bones found in piles (Yalçınkaya, Taşkıran, Kösem, Özçelik, & Atıcı, 2002). As a result of the preliminary studies on the animal remains found in Chamber E, these species were determined: Hippopotamus (*Hippopotamus amphibius*), gaur (*Bos primigenius*), wild horse (*Equus* sp.), red deer (*Cervus elaphus*), Anatolian fallow deer (*Dama dama*), bezoar goat (*Capra aegagrus*), wild sheep (*Ovis orientalis*), wild boar (*Sus scrofa*), cave bear (*Ursus spelaesus*), panther (*Pantera pardus*), red fox (*Vulpes vulpes*), and the tortoise (*Testudo* sp.). Other than these, birds, rodents, and crabs are represented in general categories (Yalçınkaya et al., 2002).

4.2.22 Keş Sinkhole

Döngel–Kahramanmaraş
Total Length: 1801 m
Total Depth: −728 m

Keş Sinkhole resides in the Keş Mountain north of Kahramanmaraş, at an altitude of 1800 m. A long penetration could not be procured during a diving activity in Yeşilgöz Sinkhole, a "Vauclusian Spring" which resides at the skirts of the Keş Mountain at an altitude of 680 m, due to the tightness of the underwater passages. The area was studied, and it was discovered that the source feeding this spring is another sinkhole at a high altitude and Keş Sinkhole was found (Fig. 4.30).

Fig. 4.30 Keş Sinkhole (*Photo* Ç. Çankırılı)

Even if this region known as Eastern Taurus Range is mostly composed of limestone and dolomite, there are at least five different formations dated to different periods. These sediments, which are mostly a product of a deep nautical environment, are thought to have been reformed through tectonic movements and embarkments, gaining a complex appearance. East and west of this 15 km^2 area, which completely belongs to an orogenic belt, are geologically distinct from one another. Even if the widest rock formation seen on both sides is limestone, there are great age differences among these limestones belonging to five different formations (Gül, 2000). Hacıveliler Formation can be observed toward the east and southeast of this area. This formation, which embodies Yeşilgöz Sinkhole and many aquifers at low altitudes, is composed of light-colored and coral foraminiferal reef limestone, rich with fossils. There are two different views considering the age of this formation: Early–Mid Miocene and Late Miocene (Gül, Darbaş, & Gürbüz, 2005). Permian dated gray, dark-gray and re-crystallized Keşdağı Formation resides northeast of this formation, which includes dolomitic limestones in some parts (Yümün & Kiliç, 2002). Different formations in the area present a highly complex hydrogeological appearance. Because there were no dye tests conducted in the eight different sinkholes and 11 aquifers found within this 15 km^2 area, the origins of these springs could not be determined. One of the most typical examples of this situation is the Keş Sinkhole.

Two small creeks unite and sink to this sinkhole that resides 1100 m above the Yeşilgöz Sinkhole and at 1800 m altitude. Geological formation in the sinkhole changes at −175 m to a vertical fault mirror, which lowers down to this point with small descents. After this sheer single drop of 170 m, sinkhole continues with small descents again and ends with a narrow siphon at −728 m. There is a 4 km distance and 400 m difference in height, between this ending point of the cave and the Yeşilgöz Sinkhole, where the water coming from the Keş Sinkhole is believed to resurface. The study of Keş Sinkhole is conducted by OBRUK Cave Research Group between 2009 and 2012 (M. Eğrikavuk, 2013).

4.2.23 Kırkgözler Cave

Döşemealtı–Antalya
Total Length: 821 m
Total Depth: −83 m

Kırkgözler Cave is a big aquifer of west Taurus Mountains, which is 35 km north of Antalya and 14 km north of its Döşemealtı county. The Taurus Mountains, which stretch out in the north–south direction in the area, acceleratingly lowers down to Antalya travertine plateau. At the contact point of these two different formations, there are many caves in Taurus Mountain's Mesozoic aged limestones. There are also many aquifers besides these caves which of many have been inhabited in prehistoric ages and have archaeological importance. Those aquifers have very low intercrystalline permeability where the limestones and dolomites comprising the aquifer have been deeply buried and deformed, destroying the original porosity. Groundwater flowing through the aquifers, dissolved conduits, and caves. In the Kırkgözler region, groundwater from the aquifers discharges at the contact point with the Antalya Plateau where the water either flows over land or is diverted underground, then discharges at secondary springs in the Antalya travertine aquifer further down gradient. Further south along the Mediterranean coast where the Taurus Mountains border the sea, groundwater from the Taurus Mountain aquifer discharges at springs near or below sea level. Current hydrogeological interest is focused on developing groundwater resources from the Taurus Mountain aquifer that either discharges to the Mediterranean Sea or leak into the Antalya travertine aquifer. The Antalya travertine aquifer consists of flat-lying permeable Quaternary travertine deposits that extend for approximately 800 km^2 and constitute the Antalya Plateau. The travertine plateau is divided into three levels where the total thickness reaches 150 m. The uppermost level has an average elevation of 250–300 m and constitutes the catchment area for the Kırkgözler springs. The combined discharge from the Kırkgözler springs supplies water to Antalya either directly or by recharging the Antalya travertine aquifers.

The biggest of the three underwater caves being researched in the area is Kırkgöz–Suluin Cave System (Fig. 4.31). The other two underwater caves, namely Kirkgoz 1 and 2 springs, have shorter penetration distances. There have been

Fig. 4.31 Entrance of Kırkgözler Cave (*Photo* G. Ture). The stalactites and the travertine formations are the proofs that the cave was in a vadose zone for a long time

ongoing archaeological excavations for many years in Kırkgöz–Suluin Cave's dry parts, and this vadose part of the cave is going to be explained separately. Furthermore, research of the underwater part of this cave, which resides 15 m below the vadose part of the cave and is reached through a chimney, concluded with the exploration of the longest known underwater cave in Turkey. An exploration into Kırkgoz–Suluin underwater system revealed a permeability structure dominated by extremely large chambers resembling megascopic vuggy porosity. The huge chambers were interconnected by small passages that circumvented large breakdown piles. The largest underwater chamber discovered was measured 100 × 60 × 45 m (Kincaid & Jablonski, 1996). That chamber was named "Stadium" and the survey shot along the west wall measured over 100 m in a straight line. Survey data from the stadium indicates that it is one of the largest known underwater chambers in the world (Fig. 4.32). A total of 821 m of conduits were explored and mapped, reaching a maximum depth of 83 m in one of the small connecting passages. The average water depth in Kırkgöz–Suluin was approximately 40 m. Water clarity was excellent in which visibility reached over 30 m in the large rooms and was only limited by the diver's lights. Percolation from the ceiling reduced visibility on repeated dives. Large speleothems were encountered down to approximately 50 m below the water surface, suggesting that the water table has been significantly lower in the geologic past. Inspection of the cave entrance indicates that at least part of Kırkgöz–Suluin is aligned with a geologic structure, most likely a small fault. However, the identification of major geologic structures inside the conduits was problematic. Large continuous conduits were not discovered. Instead, exploration and mapping revealed several large chambers connected by small conduits that circumvented large breakdown or debris piles.

Much smaller and shallower conduit systems were encountered in Kırkgöz 1 and 2 springs. Both systems were explored to the end of all apparent passable conduits but no connections to Kırkgöz–Suluin were found. Maximum penetration did not exceed 200 m, and the water depth did not exceed 20 m. The average conduit diameter was only two meters. The water clarity in the conduits was excellent. No speleothems were encountered. These facts indicate that the springs discharging along the junction between the

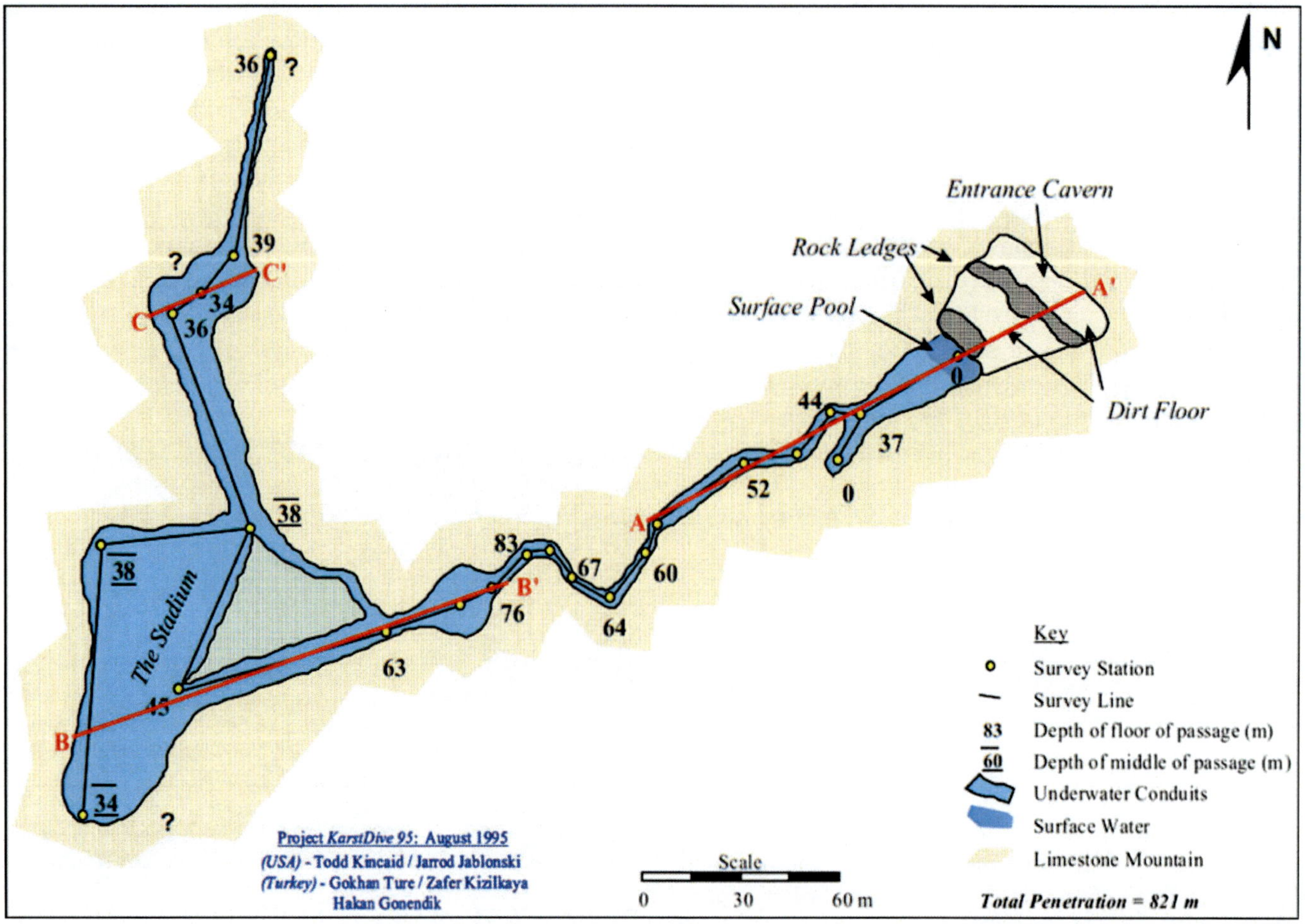

Fig. 4.32 Plan of Kırkgözler Cave adapted from (Kincaid & Jablonski, 1996)

Taurus Mountains and the Antalya Plateau represent new dissolution paths that have adjusted to the current base level. The aforementioned three underwater caves have previously been dived into a few times, but the detailed survey this text is based on has been conducted by Todd R. Kincaid and Jarrod Jablonski in 1995, during Project Karst Dive (Kincaid & Jablonski, 1996).

4.2.24 Kızılelma Cave

Ayiçi–Zonguldak
Total Length: 6630 m
Total Depth: −114 m

Kızılelma Cave is located near to Ayiçi Village at the Gelik region of Zonguldak. The cave has been developed within Lower Cretaceous aged limestones, and highly cracked forms of these limestones have increased karstification. Again Lower Cretaceous aged impermeable layer found below this limestone is in a position of a karstic ground level. Faults found within the area were effective on the formation and development of karstic forms on the surface along with the elongation of the caves. Kızılelma cave has two entrances. The main entrance is on the eastern slope of the Kızılelma uvala. This entrance found 5 m above the uvala floor is on a visible faultline and is completely fossilized. The second entrance is a −85 m deep sinkhole found 1.5 km north of the main entrance and lowers down to the middle of the cave. The first part of the cave that reaches out until the side branch toward the east, developed in the southwest–northeast direction. Debris, sand, and gravel found within this area which has four different water input points cover a large area. Pools and ponds of different sizes can be found among these. Surface waters coming to the Kızılelma uvala do not enter the cave directly. They flow underground through dolines found in front of the entrance. These dolines cannot drain the waters coming to the catchment during the rainy seasons, turning the uvala into a lake. These waters going underground through the dolines join the cave after 250 m. A small part of the uvala waters enters the Kızılelma Cave through a small creek, and this 400 m part found just after the entrance is the area where the Kızılelma Cave ceiling height is at its lowest. The ground of this section, which has approximately 50 cm ceiling height and developed according to the limestone layer direction, is covered with large gravel and sand. The cave heads toward northwest following the junction point of the main gallery and the side branch coming from east. The ceiling height of this section, where the amount of flowing water increases considerably, is very high compared to the first section and reaches 20 m in some parts (Nazik et al., 1995). There are two big collapse halls inside Kızılelma Cave; one in front of the natural chimney, one 250 m before the siphon at the end of the cave. An underground creek flows under the large rock blocks that cover the ground in these chambers. Long and deep lakes formed between rock piles. Kızılelma Cave ends with a tight siphon −114 m lower from the entrance height. Lake formations appear in front of the siphon during the rainy seasons. Waters sinking in this siphon re-appear in Cumayani Cave after approximately 250 m later. Kızılelma Cave, which connects many karstic closed basins underground, has the property of showing development in two periods. The cave is two floors in some parts, which is a sign of these periods. The upper floor is 25 m above the main gallery which can be seen from the junction point of the main gallery and the large side branch is filled with fossilized sand and gravel. There is a difference of 132 m in height between the Kızılelma Cave entrance and the mouth where Cumayanı Cave waters appear. This difference is caused by the declivity of the formation the cave formed within, and Cumayanı Cave carries the same properties. Ground formation creates an impermeability curtain for limestones and even the faultlines in the area could not change this hydrological property. Even though many faultlines are cutting the cave, there is not any loss of water. On the other hand, big collapses, halls, side branches, and water input points appeared at the parts coinciding with these faults. Kızılelma Cave, which drains the surface waters of an approximately 45–50 km^2 area, is the main vein of this area hydrologically. All the waters flowing underground through sinkholes and dolines in the area gather within Kızılelma Cave. These waters enter the cave from nine different points depending on the visible cracks and faults. Waters, which have a significantly increased flow rate at the rainy seasons, enter Kızılelma Cave without going through any apparent filtration and resurface at Cumayanı Cave. Unfortunately, signs of pollution can be seen within these waters polluted by various coal mines found in the region, in both caves. Kızılelma Cave is covered with all kinds of dripstones; some stalactites, stalagmites, columns, and curtains reached large sizes (Güngör & Gülez, 1992). On the other hand, these formations can only be seen where the underground water erosion is low, in galleries with high ceilings, on sidewalls of fossil floors and cave formations. The cave was first surveyed by cavers from Trent Polytechnic University in 1978 (Watkins, 1980). It was researched and surveyed again in 1995 by MTA (Fig. 4.33) (Nazik et al., 1995).

4.2.25 Kızılin Cave

Burdur
Total Length: 2100 m

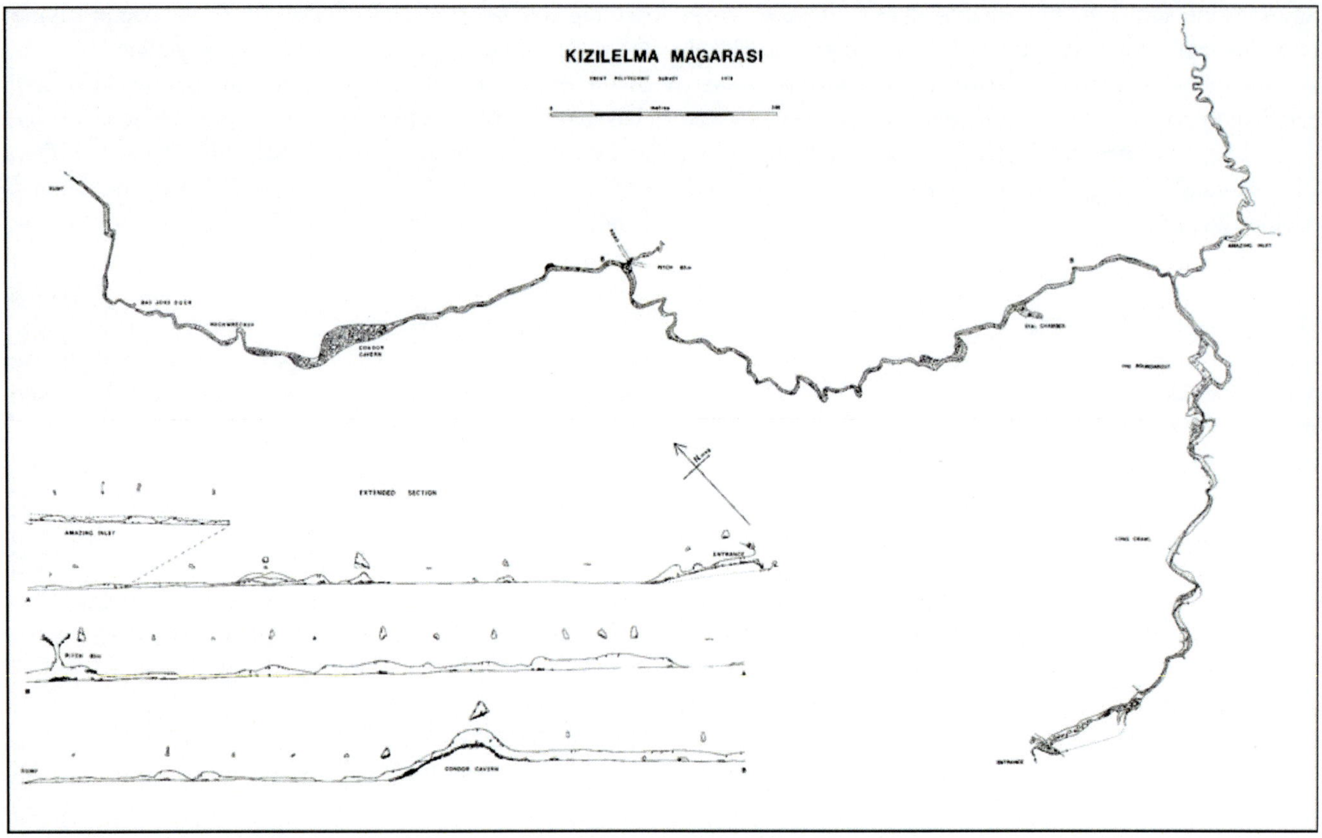

Fig. 4.33 Plan and longitudinal profile of Kızılelma Cave adapted from (Watkins, 1980)

Kızılin Cave is 13 km away from Burdur city center and 2 km north of Insuyu Cave. The geological formation of the area where the cave resides is almost completely the same as the Insuyu Cave's and structurally very complex. There is a dense, Early Paleocene aged melange cluster on top of an area formed by Upper Cretaceous aged two different formations, which divergently outcropped on top of each other (Karaman, 1994). Limestones composing the Kızılin Cave are neritic, having large pores and many cracks. On the other hand, N–S oriented active faults in the area caused the formation of Kızılin Cave as well as the collapse of many parts of it including its entrance. Following the first studies, the entrance of the cave, which resided 70 m above the plains, collapsed during an earthquake and was dug open before the 2007 exploration. The total survey length of the cave is 2100 m but this length is measured because of the halls formed on top of each other and maze-like galleries. The main axis length of the cave is just 200 m. Continuation of the big collapse hall just below the entrance has halls, which are connected through narrow passages. This section continues until the underground water basin and reaches lakes. A serious decline in water amount in both the Kızılin Cave and Insuyu Cave is observed in the recent years, because of the increasing water drainage from the plains through artesian wells, due to agricultural needs (Erdogan et al., 2014).

Chambers above lake lead to the Big Chamber, which has a size of approximately 60 × 30 m and a height of 5–10 m, following a narrow passage toward east. This chamber caused by a north–south oriented fault line has countless stalactites, stalagmites, and speleothems. There is another lake in the northeast direction of the Big Chamber (Fig. 4.34).

Most interesting findings of Kızılin Cave were two intact potteries found around this lake and human skeletons, pottery shards, artificial walls and fireplaces found in different points of a chamber that resides at the upper level, southwest of the Big Chamber. These rich archaeological findings found in this very hard-to-reach point prove that there should be another entrance at this part of the Kızılin Cave. After further research, it was understood that there truly was another entrance closed down by a collapse (Döker et al., nd). The collapse of this entrance caused the preservation of all these archaeological findings until today. All the pottery and skeletons in the cave were collected by Burdur Museum archaeologists in 2010. Even though the skeleton remains, pots and broken pot pieces found in this part of the Kızılin Cave led this place to be considered as a necropolis, a variety of pots and the burn marks on the pots later strengthened the idea that the cave was used as a living space. According to optical stimulated luminescence (OSL) tests conducted on three different pottery examples, these ceramic pieces were

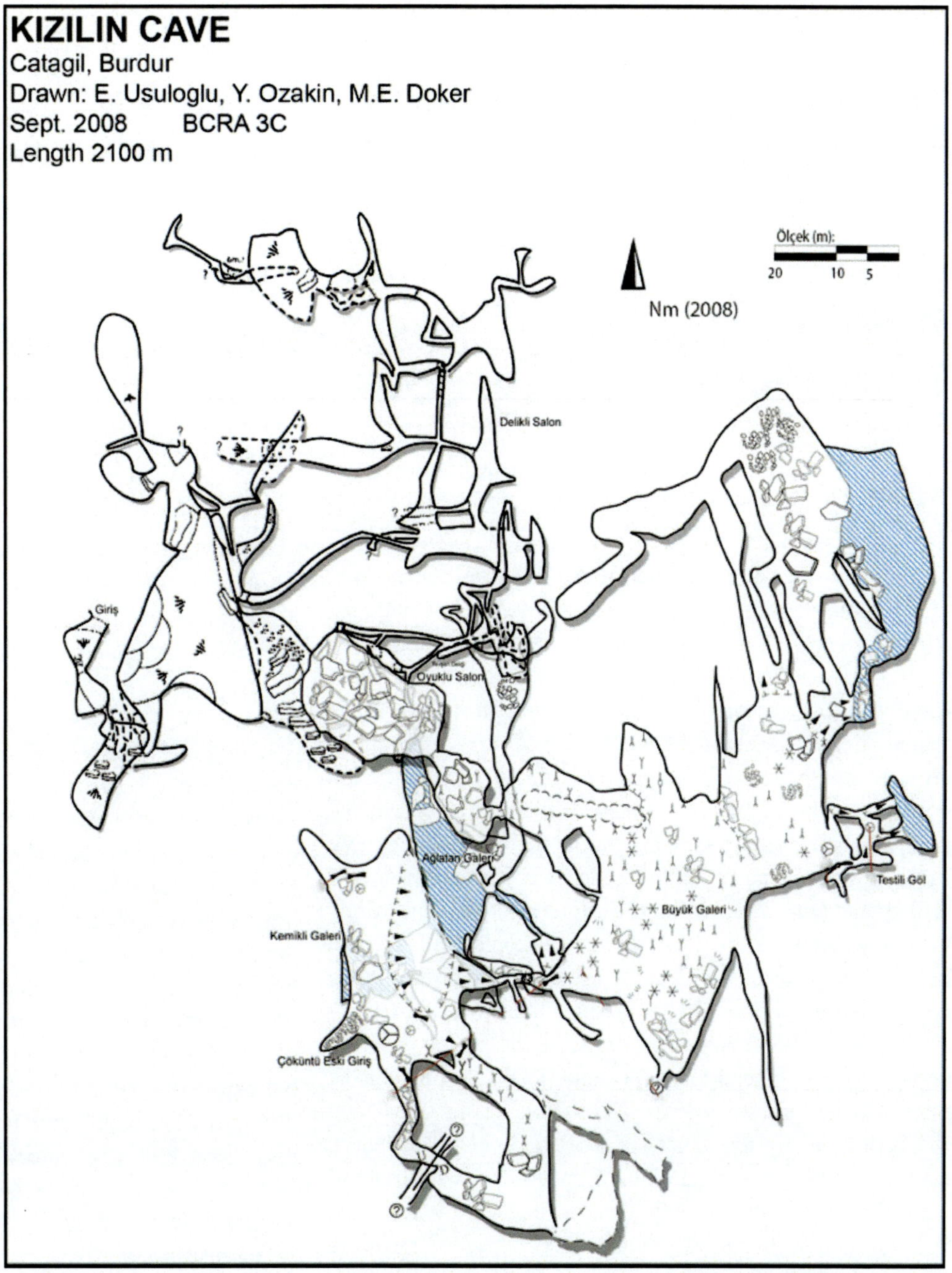

Fig. 4.34 Map of Kızılin Cave (Döker, Özakın, Gürcan, & Erdoğu, nd)

dated to BC 1443 ± 275, BC 1725 ± 336 and BC 904 ± 240 (Yasar, Demirel, & Cankaya, 2012).

There are small horseshoe bats at the entrance and around the lake northeast of the Big Chamber with troglobite isopods found at some spots of the Big Chamber. Kızılin Cave was explored and surveyed by BUMAD in 2007–2008.

4.2.26 Kocain Cave

Ahırtaş–Antalya
Total Length: 669 m
Total Depth: −79 m

Kocain Cave is close to the Ahırtaş Village; 45 km north of Antalya, located on the Indağı with an altitude of 1171 m. Western Taurus Mountains rises like a wall after the travertine plains north of Antalya and it is made up of Mesozoic–Tertiary comprehensive limestones in this area but Lower Miocene limestone lines can also be observed toward both its west and its east. Kocain Cave formed within small-grained and massive Mesozoic limestones as a result of a very big depression inside the mountain. It has the largest chamber (37,000 m^2) and the largest cave entrance in Turkey. It consists of two connected large chambers, and the width of its entrance is 65 m. The ground of the first chamber at its entrance is completely flat, but the rest of the

Fig. 4.35 Kocain Cave, the height of the column on the back wall is 27 m (*Photo* A.E. Keskin)

cave is beveled because of the rocks fell from the ceiling. The heights of some stalagmites at the end of the entrance chamber reach 16 m, and the height of the column at the second part is 27 m (Fig. 4.35) (Gilli, 1984, 1986). Kocain Cave was first explored by Guiseppe Moretti in 1919. The Italian archaeologists, who were invited for explorations after the occupation of Antalya and its vicinity by the Italian government based on the Armistice of Mudros, conducted comprehensive surveys in the entire region. During these surveys which contributed to the discovery of many ancient settlements until 1921, Moretti explored Kocain Cave, drew a highly accurate map of the cave, copied all Roman inscriptions, and drew a detailed sketch of the cistern dated back to Roman Perion in a very short time (Moretti, 1923–1924b). After 100 years, some of the 28 inscriptions Moretti copied cannot be read today (Yamaç, 1990). Following the map Moretti drew, Kocain Cave was re-surveyed and mapped again by the cavers of SCP in 1977 (C. Chabert, Callot, Chabert, & Gilli, 1978; J. Chabert, 1979a). The third map, C. Chabert started in 2001, was finalized by Ezgi Tok in 2013 and is the most comprehensive map of the cave (Fig. 4.36). Even though there was no detailed archaeological research conducted in the cave until today, the inscriptions Moretti explored were subject to various studies. There is a limestone rock mass inside the cave close to the entrance suitable for writing on, with names of many *eirenarkhes* and their acolytes *diogmites* inscribed on it. Moretti commented about the *eirenarkhes* title written on many of these inscriptions and proposed that they belonged to representatives of *eirenarkhes* who were sent on a mission in the area to participate in the holy ceremonies carried out in the cave in different periods. It is thought that these officials encountered in many inscriptions were established imitating the law enforcement in Hellenistic Ptolemaioses of the Roman Empire Age. It is known that *eirenarkhes* mean "leader of peace," and their mission is to oversee the application of customs, sustain social discipline, and prosecute bandits. SCP member Yann Callot, who was on the second team that studied the cave scientifically in 1978, accepts the *eirenarkhes* title to be some sort of local law enforcement and concludes that Kocain was a vantage point for a nearby site. In his articles, Öztürk studied these inscriptions in detail and dated all of them before AD 2nd Century (Öztürk, 2015, 2018).

4.2.27 Körükini and Suluin Caves

Çamlık–Derebucak–Konya
Total Length: 1936 m
Total Depth: −107 m

Körükini and Suluin caves are located 500 m southwest of the Çamlık Village of Derebucak District of Konya, 1.5 km far from Balatini Cave. Uzunsu River, which drains the waters of this cave, flows through Körükini and Suluin caves and resurfaces in front of the Balatini Cave (Fig. 4.37).

Körükini, Suluin, and six other caves and sinkholes nearby have very similar geologic and geomorphologic properties. They were all developed within the Jura–Cretaceous limestones of the Beysehir–Hoyran Nappes. Surrounded by impermeable units, these limestones are quite thin in the area. Triassic marns and Periman dolomites under the limestones present a ground level for karstification. Even though Körükini and Suluin Caves are separate, and they are apart of the same underground system. These dual caving systems, which mainly elongated in the east–west direction, have been diverted by the depression doline in between. It is thought to be highly possible that this cave system, which was once a single phreatic system, formed after Upper Pliocene and was divided by the doline collapsed in this same period (Nazik et al., 1993).

Uzunsu River, which originates from the east of Çamlık Village, flows through deep gorges and canyons until it reaches the Körükini Cave. The river that enters Körükini Cave southeast of Çamlık resurfaces at a depression doline after 1330 m. Width of the Körükini Cave is between 8 and 15 m, and its height changes between 4 and 20 m. Even though it is obvious that the cave, which has many lakes of different sizes inside, was formed through physical abrasion rather than dissolution, there are dripstone decorations where overflowing waters of the underground creek cannot reach to. Sand and pebble islands appear between these lakes during the low season. The Uzunsu River outflowing from Körükini Cave flows through the rocks along the 315 m long collapsed doline and enters into the Suluin Cave. This second cave is 291 m in length, 5–15 m in width, and 10–15 m in height. Most of the Suluin Cave, approximately 200 m of

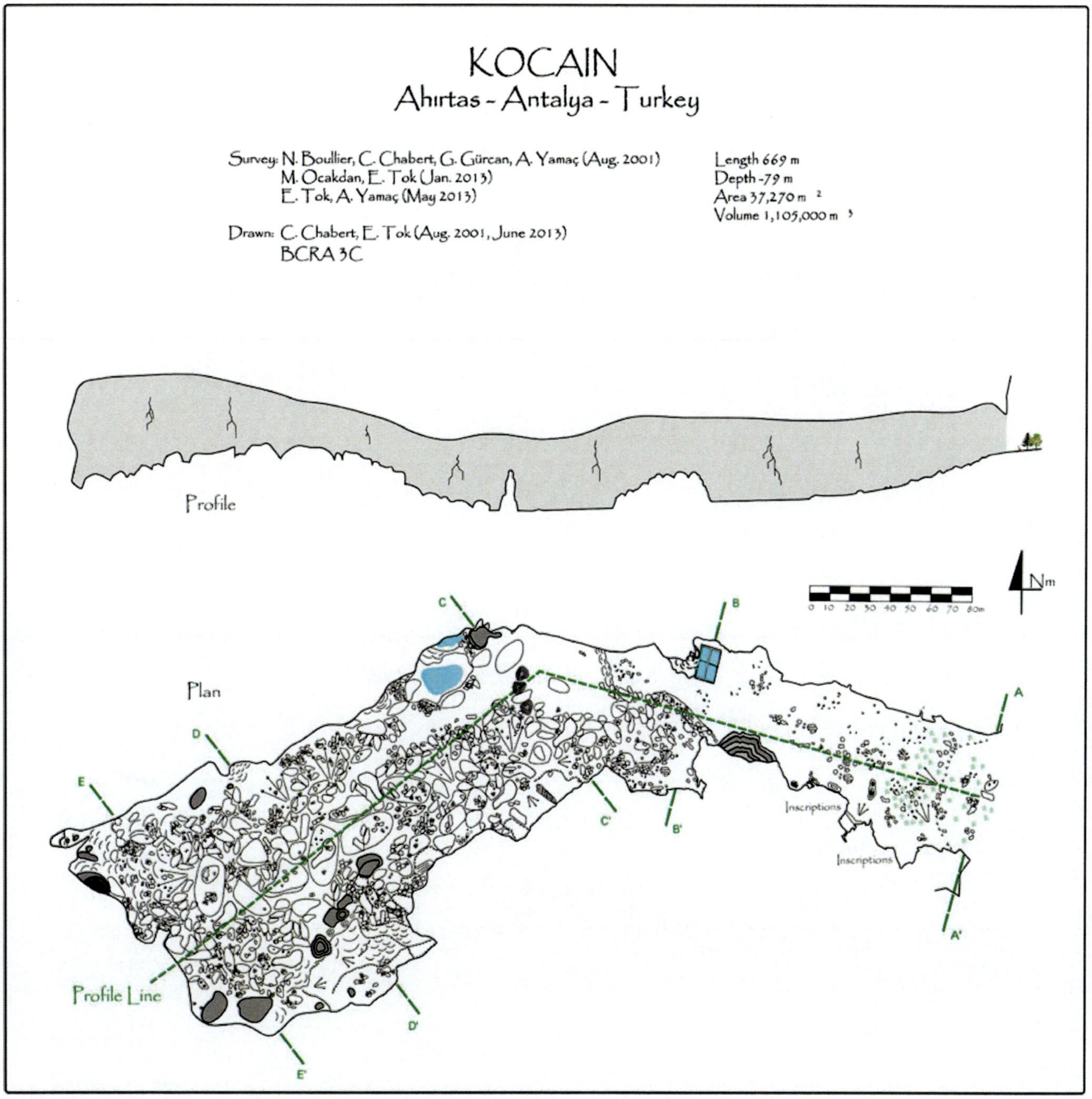

Fig. 4.36 Profile and plan of Kocain Cave drawn by C. Chabert and E. Tok (Yamaç, 2013c)

it, is a big lake. The cave, which has −107 m difference in height between its starting point and its endpoint, is 1936 m long in total (Fig. 4.38). After exiting Suluin Cave, Uzunsu River continues to flow in deep canyons and reaches the Kembos Polje. It goes underground again through two big sinkholes west of the Kembos Polje and resurfaces in the Altınbeşik Cave 35 km south, which has been proven through a series of tracer tests (Aygen, 1984). Körükini and Suluin Cave system, which was first surveyed and mapped by T. Aygen and the SCP in 1966 was surveyed in detail by MTA in 1992.

4.2.28 Kuzgun Cave

Aladağ Mountains–Niğde
Total Length: 3187 m
Total Depth: −1400 m

Kuzgun Cave is found in the Aladağlar Mountain Range, which elongates along the north of the Niğde Province, at 2840 m altitude. The southeast part of Turkey is an active plate boundary where the Arabian and the Eurasian plates are colliding. The tectonics of the Aladağlar area is

Fig. 4.37 Exit of Körükini–Suluin Cave System (*Photo* A. Aslan)

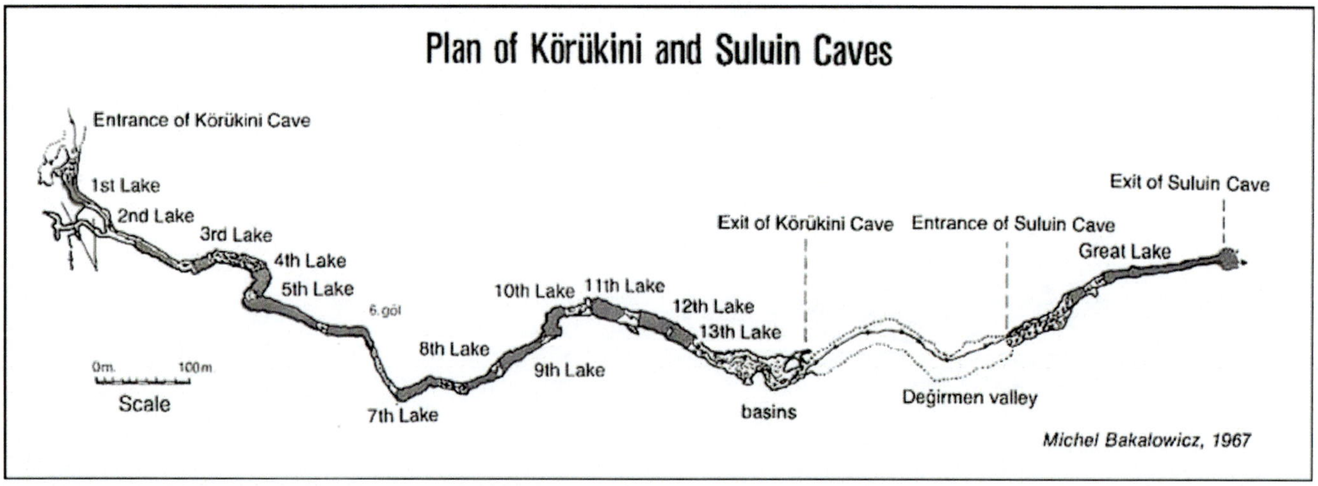

Fig. 4.38 Plan of Körükini–Suluin Cave System (Bakalowicz, 1967)

dominated by a then appes structure. During Late Cretaceous, pervasive SE to SSE folding occurred within Mesozoic carbonate platform units, associated with southward emplacement of a regionally extensive ophiolite. Autochtonous carbonate formations of Jurassic to Cretaceous age crop out in the eastern part of the area. The main part of the Aladağ Massif is allochthonous, composed by Triassic limestones, the largest carbonate nappe units. A nappe consisting of ophiolite mélange conceals the boundary between the allochthonous and autochthonous units along the eastern side. By Late Miocene, the ophiolitic cover had been removed over considerable areas, revealing Mesozoic carbonate rocks of Aladağlar. The regional evolution is believed to be closely guided by the Ecemiş Fault Zone that borders Aladağlar on the west, the site of 60 km of lateral strike-slip movement. The present-day morphology of the Aladağlar massif and surrounding basins is established during Pliocene (Tekeli, Aksay, Urgun, & Isik, 1983). The subsequent geomorphological development in Plio-Quaternary time, as well as the development of karst systems, was strongly affected by the intense uplift of the massif, downcutting of rivers, and by the fluctuations of glaciers in the Aladağlar Massif. The high-altitude part of the Aladağlar massif has been severely glaciated during Pleistocene. Glacial erosion was the dominating factor in the overall surface morphology development, resulting in the formation of numerous glacial valleys, circuses, narrow jugged ridges, and pyramidal peaks (Bayarı et al., 2003). Although glacial landforms indicate the existence of numerous episodes of glacial advances and retreats, evidence of older glaciations is largely erased by the effects of the last remarkably extensive glaciation. Common glacial valleys extend from source areas at 3100–3300 m to altitudes of about 1900–2300 m, although some large valleys cut as deep as up to 1100 m elevation. Glacial geomorphic processes on the karstified limestone substratum gave rise to distinct peculiar features known as glaciokarstic morphology. In the valley bottoms, large and deep (up to 100 m) closed glaciokarstic depressions are common, which drained subglacial flow into the karst system (Bayarı et al., 2003; A Klimchouk, Bayari, Nazik, & Törk, 2006).

The area of the remarkable karst landscape, resembling polygonal karst, with numerous steep-walled depressions, pits, and deep karrens, lies along the eastern flanks of the massif, making a ledge at elevations of 1700–2300 m. The area of polygonal karst morphology coincides with the belt of Cretaceous limestones stretching along the Aladağ's eastern flank. The high-altitude part of Aladaglar is assumed to be the main recharge area of karst hydrologic systems that discharge as large springs at the foot of the massif. Mature karst springs are grouped in four main localities on the eastern flank of the massif, at elevations ranging between 400 and 750 m. Hydrochemical and isotopic studies suggest that hydrogeologic connection exists between the high-altitude recharge areas on the allochthonous carbonates and large springs outflowing from the autochtonous formation at the main erosion base level of Zamantı River and its tributaries (Bayarı et al., 2003).

In 2001–2004, extensive field surveys of karst and caves have been carried out in Aladağlar within the joint Turkish–Ukrainian "Aladağlar Karst and Cave Research Project." These studies resulted in the discovery and exploration of about 150 caves, including Kuzgun Cave (A Klimchouk et al., 2016). The cave had been found on a small ledge near the top of an elongated rock hill in the middle section of the Kemikli valley, at the altitude of 2840 m (Fig. 4.39). The cave was a complex structure consisting of several generations of cavities. It provided easy access to the depth of - 180 m where narrow meanders suspended the exploration. Greater dimensions of the deeper part of the cave, several effluences and tributaries left unexplored, strong air draft, and the open continuation with a large pit ahead all suggested that one of the large cave systems of Aladağlar had been eventually opened.

Kuzgun Cave has been explored and surveyed to the depth of −1400 m in the main branch and to the depth of −600 m in the Veterok branch that deviates from the main

Fig. 4.39 Entrance of Kuzgun Cave (*Photo* A. Klimchouk)

one at −480 m (Fig. 4.40). In both branches, several open leads remained unexplored. A narrow gallery at −1400 m had collapsed and blocked while trying to be opened in the following years (A Klimchouk et al., 2005; Samokhin, 2013).

4.2.29 Morca Sinkhole

Anamur–Içel
Total Length: 4068 m
Total Depth: −1210 m

Morca Sinkhole, which is almost at the center of the area known as Taşeli Plateau north of Anamur at 2150 m altitude, drains the waters of a quiet large catchment. Residing approximately 5 km east of Çukurpınar and Peynirlikönü Sinkholes, Morca Sinkhole bears similar geomorphological formations with these two other deep sinkholes. The cave is located at the contact point of Miocene and Jurassic–Cretaceous limestones in the southern part of the Taşeli Plateau. This area contains overlappings like fault, fold, and systemic joint assemblies, controlled by a tectonic line in the northwest direction. The cave, which is affected by all of the Quaternary glacial periods, continues its formation with

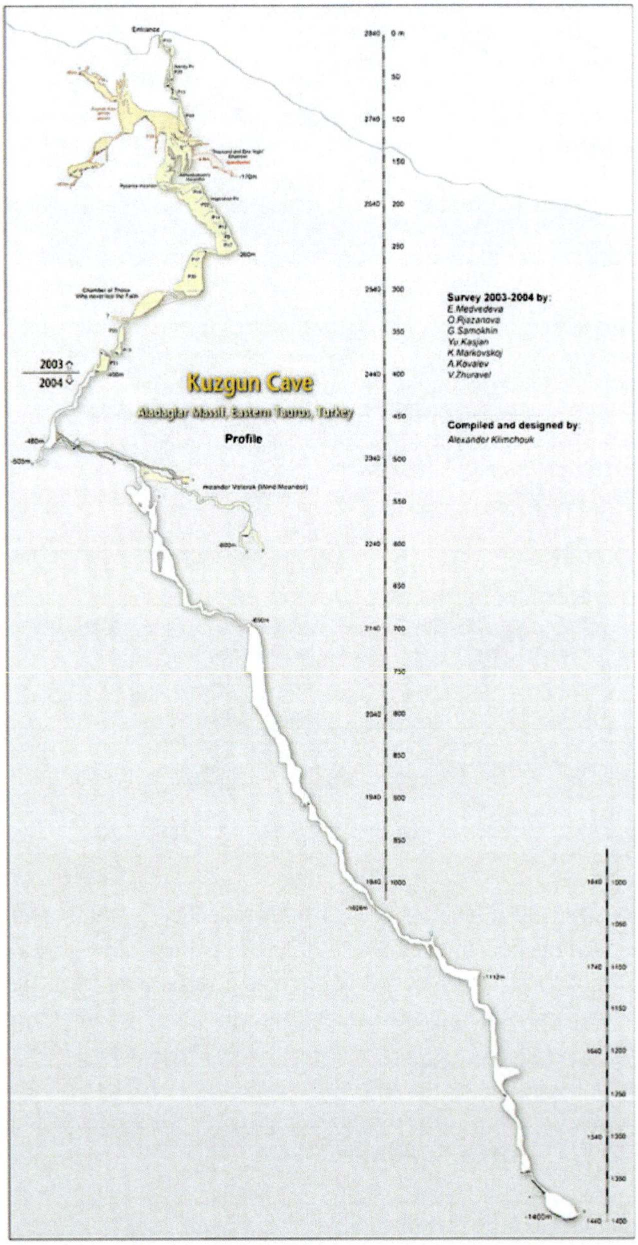

Fig. 4.40 Profile of Kuzgun Cave (A. Klimchouk et al., 2005)

snow and rainfall. Calcite veinlet and dendritic structures were observed in the foliation of argillaceous limestone, which is exposed to the compressive stress, on the shunts formed parallel to the regional tectonics and continues to form with water leaking from the potholes located southwest of the entrance of the cave. Also, a cracked surface is bearing traces of hydrothermal fluids with euhedral quartz clusters. The cave is fed continuously with water inlets at −80 m, −120 m, −1030 m, and −1040 m. Depending on rainfall and snowmelt, the siphon at the gallery at −1120 m also feeds the mainstream. The dye tests are still in the analysis process (Siler, 2016; Usuloğlu, 2019).

Fig. 4.41 Entrance of Morca Sinkhole (*Photo* E. Usuloglu)

A small side branch connects to the sinkhole, which starts with a wide entrance (Fig. 4.41) and continues with short declines, at −125 m. This active branch horizontally continues for 500 m becomes impassable tighteningly and probably carries water from another sinkhole to the main gallery. There are travertines and stalagmites in the hall at the end of this branch. The main axis of the cave continues with curving short descents without moving away from the entrance of the cave, similar to Çukurpınar and Peynirlikönü Sinkholes. The final depth reached is just 300 m north of the entrance. Morca Sinkhole was explored by ASPEG, −1210 m was reached in the sinkhole during 2019, and the exploration continues (Usuloğlu, 2019).

4.2.30 Oylat Cave

Inegöl–Bursa
Total Length: 665 m
Total Depth: +126 m

Turned into a show cave 15 years ago, Oylat Cave resides 17 km southeast of Inegöl at the exit point of the Oylat Canyon. There are Paleozoic, Permian–Triassic, Mid- and Late Miocene and Quaternary aged five different rock formations outcropping in the area around the cave. Paleozoic schists form the lowest unit. The Permian–Triassic main rock unit Oylat Cave formed within is completely made up of recrystallized calcite, and two different fault lines are crossing each other at the point where the cave is, one northeast–southwest oriented and the other east–west oriented (Nazik et al., 1997). An important amount of fractures, cracks, and fault systems developed in the unit because of the post-Permian–Triassic tectonic events. Some of these faults developed due to tectonic movements occurred before or during Mid-Miocene and some between the end of Upper Miocene and the start of Pliocene. Dating was made according to the *Gompotheriumangustiden* fossil found at the lower levels of the pile (Atabey, Nazik, & Törk, 2002). Oylat creek flowing in a deep canyon was formed at this faulty border. Oylat Cave is a fossil cave hanged at the side of this canyon. The entrance of the cave is at 525 m altitude and 5 m above the base of the canyon (Nazik et al., 1997). The cave, which has a meandering flow array, is composed of three connected parts formed in different periods. In the first part, which could be named the Entrance Hall, ceiling height is 15 m and width is 18 m (Fig. 4.42).

Karstic breccia, siltstone, and mudstone are dense in this hall. The eastern end of this hall leads outside through a small window. This window leading to the fault mirror between recrystallized calcite and Miocene sediments is the discharge point of the underground water that forms the second part of the cave. The back end of the fossil floor connects to the lower floor through a 9 m descent. Sublevel cut and grabbed the meandering water gallery and formed the fossil floor in a hanging balcony shape. Faults developed and relief system changed in the area due to the tectonic activity at the end of the Upper Miocene and the beginning of the Pliocene and thus Inegol catchment and riverbed became deeper (Törk et al., 2001). The second part of the Oylat Cave formed at that time. Previous discharge of the Oylat Cave was occurring through a chimney found 14 m above the current discharge point at first. In time, cavities formed and widened below this level through dissolution, ultimately causing the collapse of the ceiling. Following the collapse of the ceiling, the water discharging from the upper floor started to discharge from the collapsed hall. The entrance hall leads to the second part through a +4 m ascension. The second part is the meandering flow gallery with a 20 m ceiling height. There is a difference in elevation starting from +10 m and ending at +33 m compared to the entrance. There are cave breccia, stalactites, stalagmites, columns, straws, curtain dripstones, cave pearls, and dripstone pools in this middle part of the cave. The third and last part of the cave are a large collapse hall, and it is formed along a northeast–southwest direction faultline. Ground declivity reaches 40° at this part of the cave, and the ceiling height changes between 2 and 5 m. A pile with an array of pebblestone, sandstone, siltstone, and mudstone was found within this beveled part, where large blocks formed through the collapse of the ceiling. Stalactites, stalagmites, columns, and travertines are frequent at this part of the cave (Atabey, Nazik, & Törk, 2001; Nazik et al., 1997). Crumbled sediments within Oylat Cave are formed by the accumulation of sediments carried into the cave system by surface waters, while stalagmites, stalactites, cave pearls, and wall dripstones are formed by waters dripping from the ceiling of the cave. Dripstone pools are formed by the slow-flowing water stream within the cave. The cave was explored, surveyed, and mapped by MTA in 1997 (Fig. 4.43).

Fig. 4.42 Oylat Cave (*Photo* S. Çoltu)

4.2.31 Öküzini Cave

Döşemealtı–Antalya
Total Length: 60 m

Öküzini Cave lies 32 km northwest of the city of Antalya, north of the village of Yağca and 1.5 km north of Karain Cave. It is 305 m high from sea level and 5 m high from the travertine plains in front. Öküzini Cave lies within the same geologic formations with Karain Cave, the limestone of the area is mainly Cretaceous aged. Öküzini Cave is named after the ox relief on the wall of the cave (Öküzini = "Ox Cave" in Turkish) (Fig. 4.44). The incised and lightly embossed ox relief was formed as feet drawn on the lower part and head drawn on the upper part of a natural projection on the wall. The cave was named after this relief. It has been proposed that this simple relief, which was inspired from the rough shape of the rock, was once dyed. This relief is invisible due to the continuing calcification of the cave wall.

The area around the cave has an ecology with plenty of karstic waters, and small lakes form in some periods of the

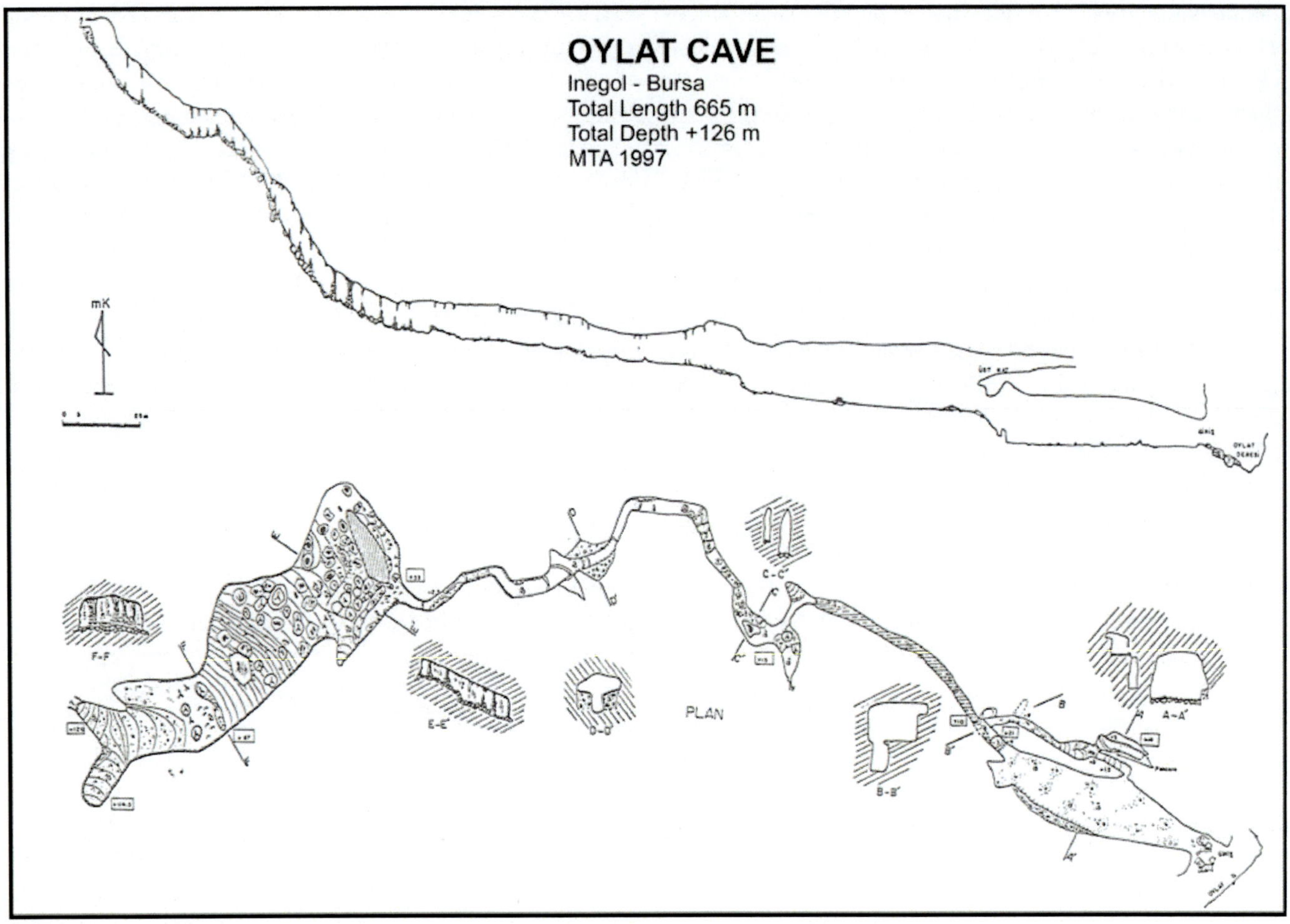

Fig. 4.43 Projected profile and map of Oylat Cave, adapted from (Nazik et al., 1997)

Fig. 4.44 Sketch of ox relief on the wall of Öküzini Cave which can not be seen today (Kökten, 1962)

year due to increased rainfall. There is groundwater in the cave much like the other cave systems in the area. These interconnected karstic water sources resurface depending on the nature of the terrain and form the natural water springs in the area. Öküzini Cave is formed through resurfaced karstic waters and became dry and habitable as the water changed its course through natural reasons like tremors. The same tremors also caused the collapse of some parts of the ceiling at the entrance. The cave essentially consists of two cavities; the big cavity where the cave opens up to the plains, which was inhabited and 23 m long, and the thin and long cavity behind this first big cavity (compartment). It faces eastward–northeastward, and it is about 25 m wide. There is a karstic water source on the ground of this dark interior cavity. There are also corridors and alcoves of different sizes in the cave. The accumulated filling inside the cave is composed of sediments that came from the natural chimney on the ceiling of the first cavity after the cave was dried out.

Öküzini Cave was first explored in 1956 by Kılıç Kökten, parallel to his Karain Cave excavations. He excavated a small test trench near the entrance. The newer and systematic excavations in the cave were conducted between 1989 and 1999 under Işın Yalçınkaya's supervision.

Besides being rich with Anatolian Epipaleolithic Age data, Öküzini Cave inhabits findings belonged to Neolithic and Chalcolithic ages in upper layers. Four different archaeologic phases were encountered in the excavations within the 3.5 m filling until today. The oldest culture Phase 1, which developed approximately between BP 17,000 and 16,500, is defined by the filling microlith tools dominated. Backed bladelets and microgravette tips are important paleolith tools of this phase. Phase 2 represents a period that inhabits non-geometric microliths just like the one below, but also geometric microliths. The dominant microlith types of this period dated between BP 15,500 and 14,200 are backed bladelets. Geometric microliths like trapezoids, isosceles triangles, and half-discs appear in this phase first. Phase 3 presents an Epipaleolithic finding group dominated by geometric microliths. This phase dated between BP 13,200 and 12,000 is the phase in which half-disk-shaped tools are the most dominant. These half-disk shapes are followed by isosceles triangles, different types of trapezoids, narrow microtips, and pointed bladelets lesser in numbers. Different sized grindstones and millstones were also gathered from these levels. On the other hand, no data on crop production was recovered from Öküzini to this date. These millstones may have been used for the milling of some wild herb products or minerals like ocher. Engravings on pebble and bone objects which bear artwork qualities are findings peculiar to this level.

Phase 4 represents the uppermost archaeological fillings of the cave. This phase, which was spoiled by Late Neolithic and Early Chalcolithic period tombs opened up inside the Epipaleolithic period layer, has the quality of a complex filling. Tombs and urns left inside them as gifts show that the cave was used as a graveyard after the Epipaleolithic period, where these last inhabitants of the cave originally lived are still unknown.

In Phase 4, several incised river pebbles were found by Kökten. One of the pebbles has a hunting scene on both faces. While one face depicts a hunting band, the other portrays a trap (Kökten, 1962). The replica of the ox or bull engraving on the cave wall is incised on another pebble. Other pebbles with what appear to be human figures have also been found. Similar incised pebbles were also found in Phase 4 of Yalçınkaya's excavation. Engraved bones, pierced seashells probably used for decoration, a necklace made from the incisors of a fallow deer, rhyolite flat and circular pierced beads, and fragments of ocher were found. Bone tools include awls of various sizes and styles, which suggests that leather working was a practiced craft. Phase 4 is dated between BP 10,000 and 7,900.

While no domestic animals are encountered among the faunal remains of the Öküzini Cave, the dominant types are wild sheep and goats. Remains of some pigs, various types of deer, and carnivores can be seen. There are many stove places in almost all phases of the cave. The researchers conclude that the stratigraphic sequence is as follows: Phase 1: Faunal assemblage includes fallow deer and goat. Phase 2: The faunal assemblage in the phase included goat and sheep. Phase 3: In addition to goat hunting, deer reenters the faunal assemblage (Yalçınkaya, 1992b; Yalçınkaya, Leotard, et al., 1997).

4.2.32 Peynirlikönü Sinkhole

Anamur–Içel
Total Length: 3118 m
Total Depth: −1429 m

Peynirlikönü Sinkhole, the deepest cave explored in Turkey, is located on the Çukurpınar Plain, close to the Olucak Village, 25 km north of the Anamur District of Içel. The sinkhole is at 1930 m altitude and 500 m far from Çukurpınar Sinkhole. This area found north of Anamur and bordered by Göksu River in the north is known as Taşeli Plateau. This karst plateau, which rises in a form of a sheer wall at some parts north of Anamur, has an average altitude of 1600–2000 m. There are numerous dolines and sinkholes on the surface of the plateau, some of which are the deepest in Turkey. On the other hand, all the waters going underground at the plateau resurface from numerous springs that form the base at 400–600 m altitude and some springs become the source for rivers with high flow rates (Nazik & Törk, 2000). In some sense, Taşeli Plateau is the largest and highest karst plateau of Turkey. Even if it looks like a high plain on a wide scale, it has a faulty structure even hard to walk on. There are rocks subsided in different conditions and settled in the Pre-Permian–Quaternary period in the area (Yapici, Anil, & Yetis, 2003). Mid-Taurus Range, especially the Taşeli Plateau, was covered by the sea in the Paleozoic. The land, which underwent an ascension through side pressure toward the end of the period, stayed above the water until the start of the Cretaceous. Trias and Jura formations also became terra-firma with this ascension and eroded afterward. With the invasion of the Cretaceous sea, Paleozoic formations became submerged again and new sediments started accumulating over. The area probably stayed as terra-firma again until the start of Lower Miocene, following the ebbing of the sea. An age of strong erosion should have begun at that time. A new sea invaded the area again through transgression abruptly. Some parts of this generally deep sea were shallow, active, and warm. After the ebbing of the sea in question through a regression, lower parts of the area were invaded by the Mid-Miocene sea through an abrupt transgression again. Clay, marl, and clayed limestone layers, which formed the Mid-Miocene formations, show us that the sea was quite deep and active at its bottom. Numerous and various fossils found tell us that both seas had suitable conditions for lifeforms (Siler, 2016). Taşeli Plateau reached its current form through tectonic movements, and these sea invasions reaching up to Mut–Karaman regions. Peynirlikönü and Çukurpınar Sinkholes reside at same altitudes, on the same fault line, with 500 m apart. There are two rock formations of different ages and properties in the area. Jura–Cretaceous aged limestone and Late Triassic period sediment rock formations border each other. This formation border, in addition to the main fault, created those sinkholes (Usuloğlu & Siler, 2015). Before the explorations of these two sinkholes, it was widely accepted that Çukurpınar and Peynirlikönü dolines were connected, both discharging the waters they drained from their catchments from Sugözü karstic spring, the main source of the Dragon Creek. Studies revealed that such a connection is not possible that there is a northward orientation underground from where these dolines are and it is far west of the Sugözü karstic spring (Siler, 2016).

Peynirlikönü Sinkhole was discovered by BÜMAK in 1990, during the exploration of Çukurpınar Sinkhole. The entrance and the first −120 m of the cave are quite tight. Following narrow galleries, the ceiling disappears (Fig. 4.45). The majority of the cave is composed of 1 m wide tunnels with high ceilings and water at the bottom. The following parts of the cave tighten down to 50 cm at some points. The "Beauty Hall" at −65 m is the only part adorned with decorations in the cave that continues with continuous descents. The lake at −232 m is the largest lake encountered in the cave. From this point to −672 m down, the cave continues with few relatively long descents. Then, narrow passages and meanders start again. A large fossil gallery along with "−700 Camp" is reached after several descents. Camping is rather difficult because the floor of the gallery is covered with fallen rocks. The water passage is very narrow at that point. The underground stream can be reached again through another descent after the fossil gallery. By the active gallery, the continuation of meanders begins again, but longer and more tiresome this time. The longest known drop of the cave is reached after the catwalks. Amount of water in the main gallery increases due to various small active branches after the descent. There are only a few flat areas down at −1429 m, where it is possible to walk just for a short distance (Fig. 4.46). Two of these areas have already been occupied as camping grounds (Döker et al., 2013; Döker, Özakın, Öztekin, & Tuncer, 2004). Many different types of amphipods and trichopoda were detected during the *biospeleological* studies conducted in Peynirlikönü Sinkhole. A new species of an amphipod was found at −650 m and was given the name "Gammarus ustaoglui" (Özbek & Güloğlu, 2005).

Fig. 4.45 One of the last descents in Peynirlikönü Sinkhole (*Photo* S. Çoltu)

4.2.33 Pınargözü Cave

Yenişarbademli–Isparta
Total Length: 8500 m
Total Depth: +440 m

This outlet cave resides in the western Taurus region on the north slope of the Dedegöl Mountain. It is 1550 m high from the sea level on the northern elongation of the Dedegöl-Dippoyraz Mountain, which has an elevation of 3000 m, 11 km west of the Yenisarbademli District of Isparta. It has characteristics of being the longest cave with the highest positive altitude of 440 m. Pınargözü Cave resides at the contact point of three different geological formations and continues along the Karagöl Formation, in the northeast–southwest direction, within Triassic aged Kasımlar Formation limestones also known as "Dippoyraz Limestones." While the main active branch, which elongates on the southwest direction, goes along a fracture line with a declivity of 60°, the fossilized branch in the east continues along with the contact point of the Karagöl Formation and the Triassic aged limestones in the southwest direction. Karagöl detritic formation has a collapsed morphology caused by the domination of marns and conglomerates. It was detected that the base level in the relatively narrow area the cave resides was composed by Karagöl and Kasımlar Formations, and the passages filled with water in the phreatic zone are situated just on the Karagöl Formations. The carbonated sequence Pınargözü Cave resides in composes the

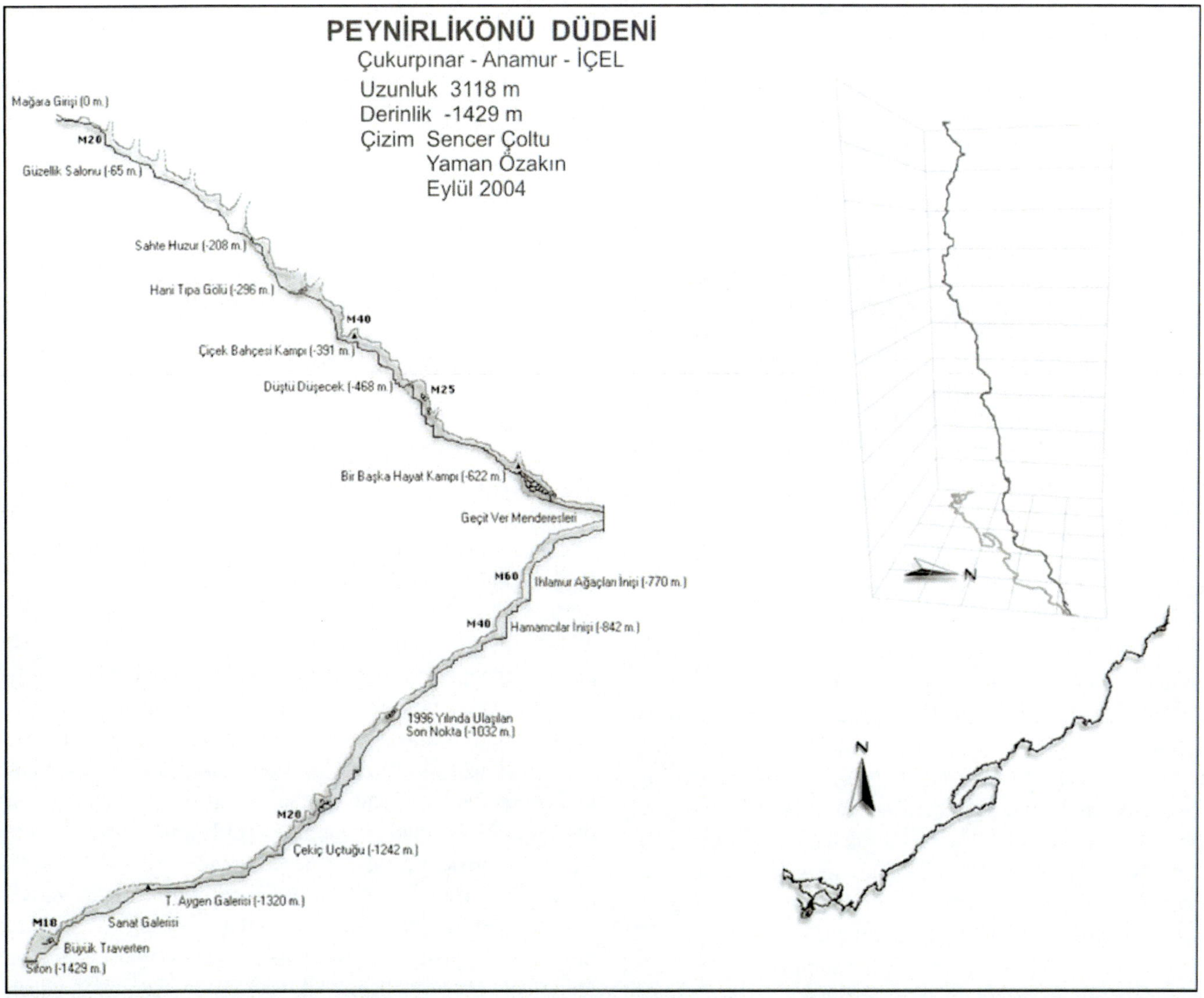

Fig. 4.46 Profile and plan of Peynirlikönü Sinkhole (Döker, Özakın, & Güloğlu, 2013)

whole top section of the massive which is more than 1000 m thick. Reef limestones are found generally at the base. Karagöl detritic formation creates a perfect guide level at the top. The last part of the sequence encompasses various carbonated facies types found in an intercalated state, which is in the group that composes the core of the massive. At least four different facies types coincide there and form a complex group very hard to separate. Dolomitic lime breccium, reef limestones, dolomites, and bituminous limestones can be observed among these. The general tabular layout of the Triassic carbonated sequence shows that the highest peaks of the massive are also stratigraphically the highest. In the northern direction, a structural dip of the carbonated sequence reveals the normal layer of this sequence: the Kasımlar Formation. This formation is completely composed of Triassic aged, gray–dark-gray-colored limestones and elongates as a layer on top of other facies until the peak of the Dedegöl Mountain (Dumont & Monod, 1976).

Water amount discharging from the entrance of the Pınargözü Cave is about 50 L/s in the high season, and the water temperature is around 10 °C even in August (Bakalowicz, 1968, 1970). The 50-m-long siphon found just after the entrance, which caused huge problems during the first years of exploration, disappeared today due to the declined water level (Fig. 4.47). Pınargözü Cave, which progresses in the form of an active underground creek until the first 1500 m and ascends to +58 m through waterfalls bearing the same characteristic properties, reaches +65 m through high berms following the hall covered with big blocks formed by dissolution (C. Chabert, 1975). Atmospheric, morphologic, and hydrologic effects caused a local climate to be formed here. Helictites are formed in the hall by the winds in the cave atmosphere caused by these local climatic effects. Following this hall, the whole cave continues with passages and difficult climbs until +440 m. Even though the winds blowing at 154 km/h measured in narrow

Fig. 4.47 Meanders in the middle passages of Pınargözü Cave (*Photo* H. Eğilmez)

galleries at +250 m during older researches (C. Chabert, 1972) slowed down, drastic factors like the 25-m-long waterfall at +350 m make Pınargözü Cave the most difficult cave to explore in Turkey.

The cave was found by Temuçin Aygen in 1964. Researches started by the members of SCP in 1968 continued until 1972. In this period, surface researches were conducted on the whole of the Dedegöl Mountain to explore the sinking points of the waters discharging from Pınargözü Cave but none of the 6–7 sinkholes were connected to the Pınargözü Cave. The length is 5275 m, and the height is +248 m in the map of the Pınargözü Cave drawn in 1977 (Fig. 4.48). Pınargözü Cave was re-surveyed by a mixed team between 2011 and 2016. Although the new map is still being prepared, it is been informed that the length of the cave reached 8500 m and its height increased to +440 m with the new branches found. A total of 50 km^2 wide area was scanned during the same research but, even though there were new caves found other than the sinkholes previously explored by the SCP, still no connection could be made to Pınargözü Cave. Thus, sources feeding this huge aquifer with a high flow rate of 50 l t/s still could not be found despite all the researches conducted for more than 50 years.

4.2.34 Sarpunalınca Cave

Devrekani–Kastamonu
Total Length: 1683 m
Total Depth: −59 m +9 m

Sarpunalınca Cave is 400 m far from Sarpunalınca Village, 22 km east of Devrekani, in the vicinity of Kastamonu, northwest of Turkey. Squeezed between the north of Devrekanı and the south of the Yaralıgöz Mountain, this region contains younger sedimentary units deposited discordantly over the Daday-Devrekanı group, the highly metamorphosed Precambrian age formation stretching to the north and the east. This formation, also called "Gürleyikdere," contains largely sedimentary and volcano-sedimentary material. On the other hand, this formation is partly integrated into the Upper Jurassic–Lower Cretaceous age with Yukarıköy Formation. These limestone outcrops surfacing in places between Şenlikpazarı and Sarpunalınca are all assumed to have been deposited in the same era. Volcanic material is commonly observed below all these limestone blocks, which

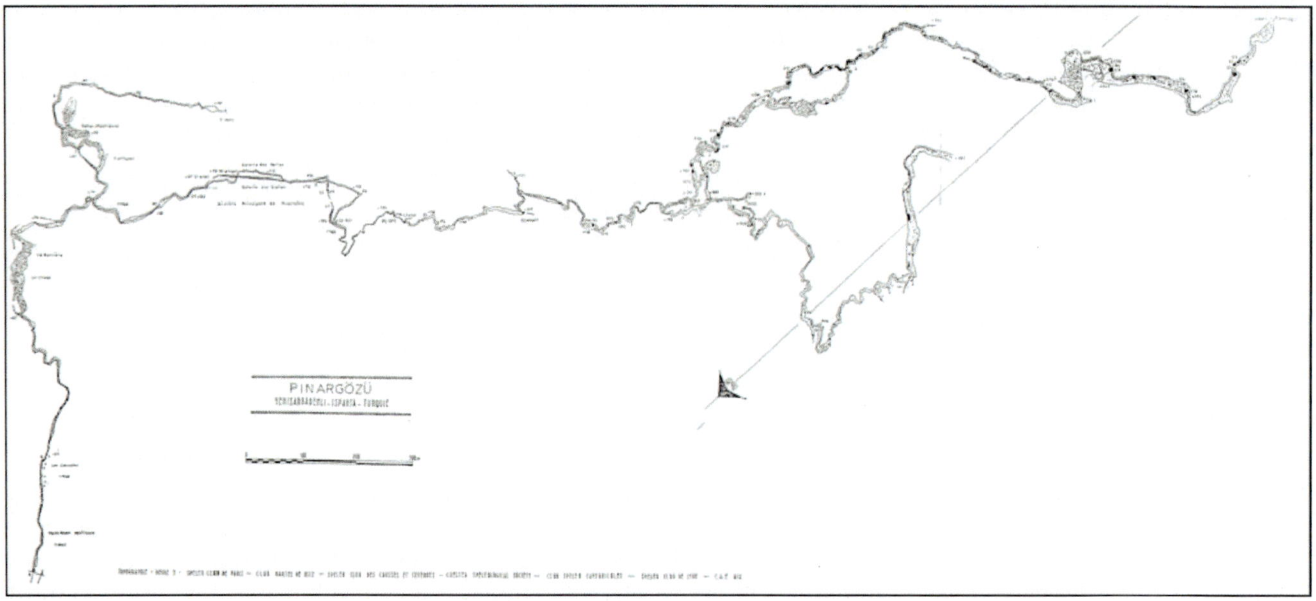

Fig. 4.48 Plan of Pınargözü Cave (C. Chabert, 1977)

are found in melanges all the way to Yaralıgöz Mountain (Tunoğlu, 1991). On the other hand, the region has at least nines faultlines within an area of 15 km^2. We can infer that the younger limestone reefs deposited over the aged underlying formation have undergone substantial dislocation and fragmentation as a result of all this tectonic activity (Boztuğ, 1992). Simultaneously, a large portion of these limestone blocks has been covered with earth. Sarpunalinca Cave was formed within the Jurassic–Cretaceous aged limestones of the area (Tunoğlu, 1991). Due to the shallowness of this limestone, and the impermeable layer below, the cave has a branching structure. The accumulation of water in the impermeable level has accelerated erosion, making Sarpunalınca an extensive system.

The cave starts with four interconnected entrances, each formed at a different age. The upper level, currently 9 m above the active stream, probably represents the location of the prior sinking stream. These entrances and the passages beyond have dried out and became fossil galleries. Over time, the second-level entrance is the large sloped passage located south of the current stream sink entrance. This wide gallery, now fossil, connects with an active stream in a large hall 13 m beyond. The third level is the lower entrance where the stream currently sinks into the cave. Following these four entrances of three different ages, Sarpunalınca Cave continues with a quite large hall where the stream flows along the southern wall 64 m into the cave, and the stream diverts to a very narrow passage on the south wall. The smaller second hall connects to the first one through a sloped passage. The combined length of the two halls is approximately 400 m and generally 40 m wide. Ceiling height reaches up to 26 m. The halls are strewn with giant blocks fallen from the ceiling, especially on the northern sides and on the eastern end. The large hall, which was the main waterway before the stream cuts through a new passage on the south, is now completely dry and decorated with stalagmites and stalactites (Fig. 4.49). This is readily observed where the stream re-emerges at the other branch and rejoins the ancient bed. The tectonic activity caused this main hall to become suspended, with the stream forming itself a new path in the cave. At the end of the halls, the cave continues along a fault, with a width not exceeding 6 m, but with ceiling heights up to 20 m. At two locations along this active passage, large fossil galleries and halls have formed on the north side, due to the seeping water from the limestone structures observable on the ground. Although partly collapsed, one of these galleries still has a connection to the surface, bringing in the water at the rainy season. This second part of the cave also contains foreign material like basalt and andesite, obviously carried over from the surface, giving an idea about the power of water flow in the rainy periods. Tree trunks and branches lodged 3–4 m above ground level are also a testament to the rising water levels.

Fig. 4.49 Sarpunalınca Cave (*Photo* Ç. Çankırılı)

Eventually, the cave reopens to the surface through multiple exits, very similar to the entrances. A little beyond the resurgence is a water mill that is known to be operational until 30 years ago. Limited biological samples were gathered. Also, a fox skeleton was found in the fossil gallery near the entrance, and an alive marten was photographed about 30 m inside one of the exits. Guano was observed in a few places, but very few bats were sighted (Yamaç, 2010; Yamaç & Yıldız, 2011).

The first exploration of the cave was carried out by BÜMAK in November 1981. In 1982, MAD visited the cave and carried out a rough survey. By 2010, Sarpunalınca Cave was re-surveyed by OBRUK Cave Research Group (Fig. 4.50).

4.2.35 Sıtmasuyu Cave

Beykonağı–Bozyazı–Mersin
Total Length: 135 m
Total Depth: −5 m

Sıtmasuyu Cave is located in an area where intense tectonism is experienced due to thrusting and nappes. The region is identified as the Middle Taurus Karst Belt, and even in a small area, several formations are observed. Permian, Jurassic, and Cretaceous carbonate rocks are overlapped with Miocene deposits. The karstification of the region is so intense that it is not a surprise to see some of the deepest caves of Turkey, namely Peynirlikönü, Morca, and Çukurpınar on the high plateau and just 12 km NW of Sıtmasuyu Cave.

The cave is on a steep valley slope deeply cut by Gökçesu stream. The entrance to the cave is at an altitude of 1200 m. Sıtmasuyu Cave was formed in the area where the cave

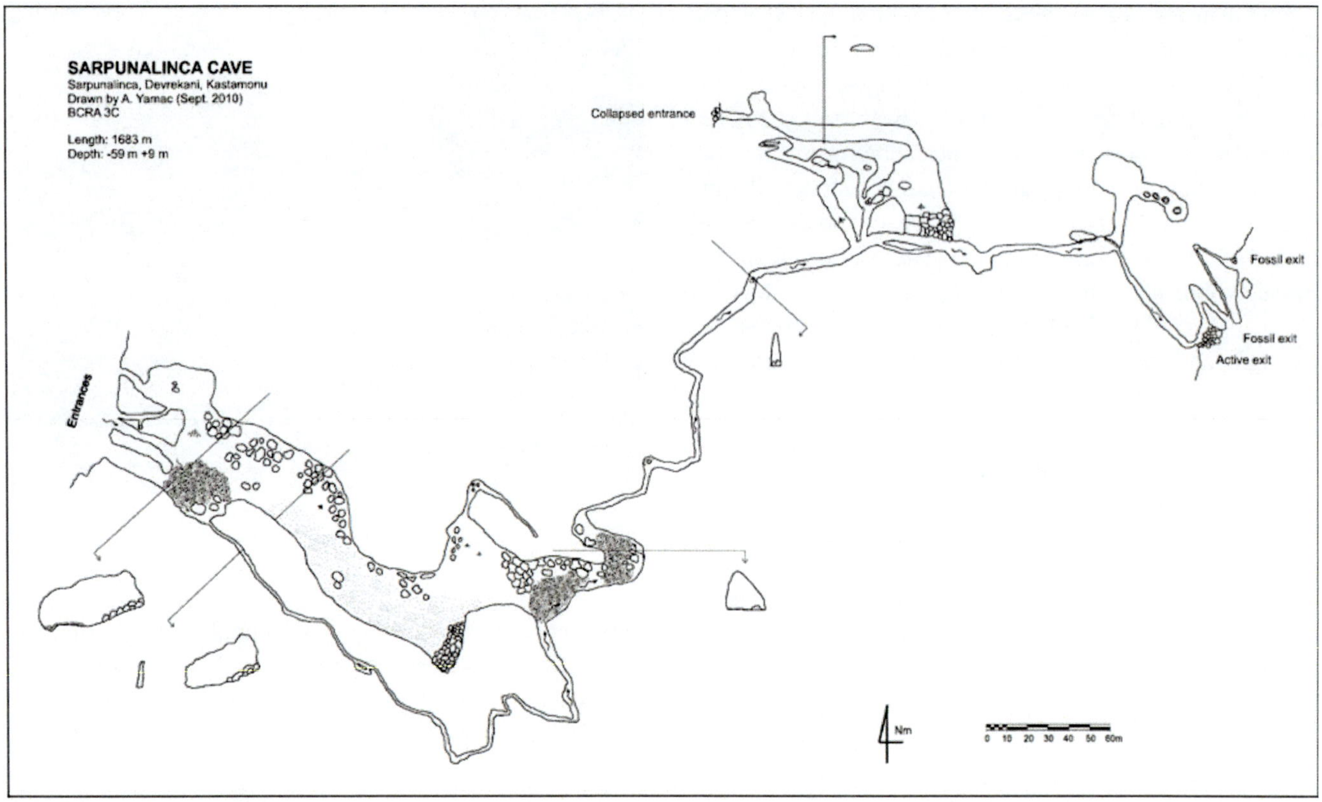

Fig. 4.50 Plan of Sarpunalınca Cave (Yamaç, 2010)

fracture system formed a shear zone in the Jurassic limestone. The dominant direction of the fracture system is NW–SE, and it is the same with the main gallery extension of the cave.

The cave, which is in the vadose zone, has small ponds with a depth of half a meter. After the entrance, the gallery descends into the widest section of the cave which is covered with boulders. Here, the cave is covered with stalactites, stalagmites, and dripstones. The last accessible section of this part is covered with columns which makes further progress impossible. The second section after the entrance gallery has two parts that are connected. Compared with the first part of the cave, this section is narrower and lower. In this last section, in addition to stalactites, stalagmites, and columns, cave sediments and shallow ponds were observed (Fig. 4.51). This section, which can have a connection with the lower branches, ends with a collapse (Törk et al., 2009).

Fig. 4.51 Last section of Sıtmasuyu Cave (*Photo* K. Törk)

4.2.36 Sofular Cave

Sofular–Zonguldak
Total Length: 490 m
Total Depth: −80 m

Sofular Cave is near to Sofular Village on the Zonguldak-Ankara motorway. The cave developed in an Early Carboniferous aged formation very suitable for karstification. This formation is limited by a Lower Cretaceous aged layer composed of carbonate cemented sandstones and an opposite fault caused a stratigraphic inversion. The Early Carboniferous aged formation the cave resides in is composed of three different layers. Cretaceous and argillaceous limestones are found at the lowest layer. While the middle layer is composed of dolomitic limestones, the top layer is composed of pure limestones. Great differences in CaC03 rates can be observed in these layers. Sofular Cave is formed in the lower floors of this formation. The lowest floor of the cave, which is at −80 m from the entrance, is formed within argillaceous limestones. On the other hand, upper floors reside within the dolomite levels of the formation. Sofular Cave resides on a plateau that is one of the most karstic catchments of the area. There are many karstic formations of different sizes on the surface of the plateau, and the most important of these formations is an old ruined polje. The base of this old polje, which developed according to the tectonic structure of the area in a southwest–northeast direction, is composed of dolines and uvalas of different sizes. Sinkholes can be seen in many of these closed formations that extend in lines. Sofular is a fossil cave with a completed hydrogeological development and a large part of it is found within the vadose zone. Specifically, the upper floors are dry aside from the waters dripping from the ceiling in the rainy seasons. On the other hand, the lowest level, where lakes of different sizes reside, is the semi-active level of the cave. Dripping and leaking waters feed this level even in dry seasons. The water level rises significantly in rainy seasons. Waters of the semi-active layer of the Sofular Cave come from the dolines found on top. Waters coming from many dolines closed to surface sink underground through sinkholes. A small part of these waters resurfaces from small springs found on the slopes of the Sofular Creek, and a large part of it reaches the lowest level of the Sofular Cave (Nazik et al., 1995). Including its side branches, the total length of the Sofular Cave is 490 m, and it is composed of three connected floors. The entrance of the cave faces the northeast, and its main gallery elongates in the east–west direction. The ground of the completely fossilized main gallery is covered with rock blocks and debris. The first part of the cave has a declivity of 45°. There is a 5 m decline following a dripstone bridge at the end of this first part. This escarpment connects to a gallery with a vertical base that develops toward the right-hand side following a −2 m descent. There is a small lake at the end of this gallery, where red soil mixed human bones are found, is found at −45 m from the entrance. The fossil gallery that continues with two steep ascents of +2 m and +6 m connects to the gallery found at −8 m from the entrance. There is a −35-m-long descent at the end of the fossil gallery. The base of the fossil gallery found at −60 m from the entrance is covered with large blocks. The waters disappear from a small siphon developed on the right-hand side of the gallery. The second and third floors of the Sofular Cave are reached through a small connection gallery that opens up toward south from the fossil gallery, and this gallery leads to the largest and the most beautiful hall of the cave. North and west edges of this hall found at −53 m from the entrance are covered with dripstones of all kinds and colors. Southeast and west edges of the hall are caved in. The cave-in at the west created a −12-m-long steep well that connects to the deepest level of the cave. This level found at -80 m from the entrance is covered with clay and alluvion piles, and it is completely submerged in the rainy seasons (Kaufmann, 1993). It was explored by MAD in 1990 (Laumanns & Özbek, 1990) and by MTA in 1995 (Fig. 4.52). The final zone of the fossil level is mainly inhabited by bats. There are also white-colored guanobia living in the guano. Millipedes and flatworms (Platyhelminthes) were seen in the entrance zone (Nazik et al., 1995).

4.2.37 Suluin Cave

Döşemealtı–Antalya
Total Length: 300 m
Total Depth: −40 m

Not to be confused with Akseki Suluin Cave, Antalya Suluin Cave is the starting point of the aforementioned Kırkgözler Cave. It is located 32 km northwest of Antalya, within the borders of Yağca Village. Suluin Cave lies 1 km northeast of Karain Cave and 125 m northwest of Öküzini Cave at an altitude of 320 m above sea level and 20 m above the plain. The cave comprised of a single chamber facing east and has a siphon lake at its end which is connected to Kırkgözler Cave. Although there is a difference in levels within the 14 m between the terrace leading to the mouth of the cave and the pool at the back of the cave, the exact thickness of archaeological deposit in between is not known. The cave roof extends eastward and appears to have collapsed at times due to earthquakes. Large calcareous blocks fallen from the ceiling are covering much of the cave's floor (Taşkıran, 2011).

The archaeological excavations started in the cave under the leadership of H. Taşkıran in 2007 are finished in 2014 because the bedrock was reached in all the pits. The archaeological fillings in Suluin Cave change between 2 and 2.5 m. There were five different geological levels detected in the excavations. All the fillings belonged to the Holocene Age while no fillings belonging to the Pleistocene period were encountered. There is a decline in and even disappearance of archaeological findings starting from the 20th archaeological level, approximately 2 m deep from the surface. It was observed that the dense reef calcareous rocks, which did not show any specific order, encountered on the surface became walls that surround some areas after 60 cm. Three different areas revealed in Suluin were named "A", "B", and "C". Among these, area "B" has a plastered floor.

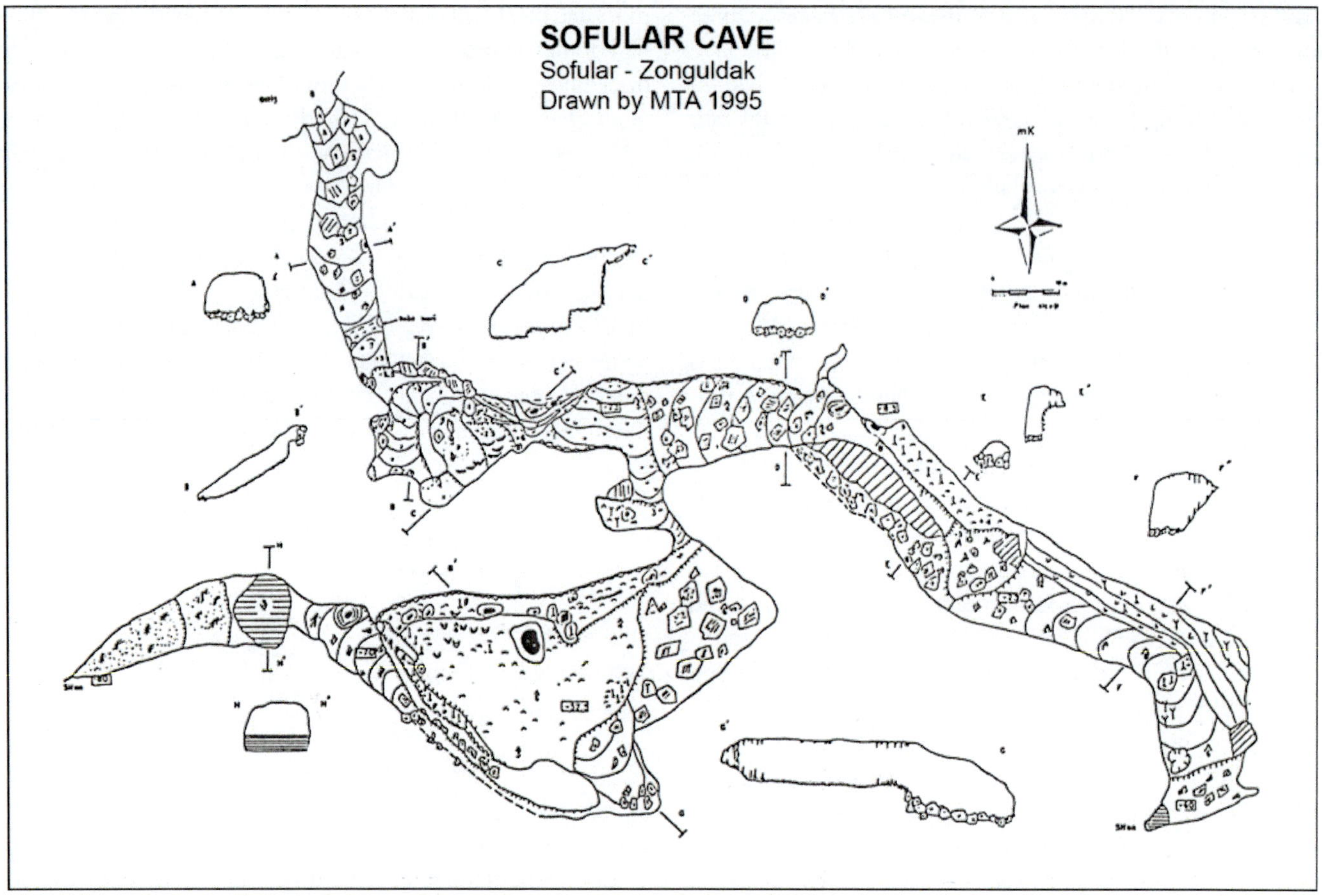

Fig. 4.52 Plan of Sofular Cave adapted from (Nazik et al., 1995)

Among the caves around Karain, including Karain, Suluin is the first cave where architectural remains were found within its layers and carry great importance because of this specialty. Excavations carried out at Suluin Cave seem to justify claims over the years that Neolithic and Chalcolithic settlements in this area should be sought in caves.

The C14 result of the charcoal samples taken from earliest archaeological levels indicates the first quarter of the 8000 BP and archaeo-geophysical studies show that Suluin Cave was not inhabited during the Pleistocene Period (Taşkıran et al., 2014). The excavations carried out in Suluin show indications of hunter–gatherers but also Neolithic, Chalcolithic, and Early Bronze Age communities who adopted a sedentary lifestyle have used the cave (Taşkıran, Kösem, Özçelik, Aydın, & Erbil, 2015).

4.2.38 Susuz Cave

Seydişehir–Konya
Total Length: 2303 m
Total Depth: −79 m

Susuz Cave lies about 1 km southwest of the Susuz Village, 15 km southeast of Seydişehir. It is also called Keş Deliği and Güvercin Taşı Deliği. Susuz Cave, which is formed within Upper Cretaceous limestones, is shaped by the dominant structures in the area, tectonic fractures and joint systems. Dye tracing tests revealed that the waters flowing into the Tinaztepe Sinkhole reach to this cave. Susuz Cave functions as an aquifer that discharges the underground water coming from Tınaztepe Sinkhole to the Suğla Lake (Sroubek, 1989). The cave has two entrances; the main entrance is located at the bottom of the valley while the second one is a wide doline developed on the slope. While the first entrance reaches the underground water with a 45° declivity, the second entrance has a −66 m steep descent. The river flowing within Susuz Cave cascades for about 70 m in a branch opposite of the valley entrance and ends with a siphon after 40–50 m. The sighting of a non-troglobite fish in front of the waterfall during the French exploration of 1971 shows that the siphon has a direct connection with the Sugla Lake and its waters rise until the waterfall during the high season (Fig. 4.53) (Raimond & Chabert, 1971). There are wide chambers at two points of

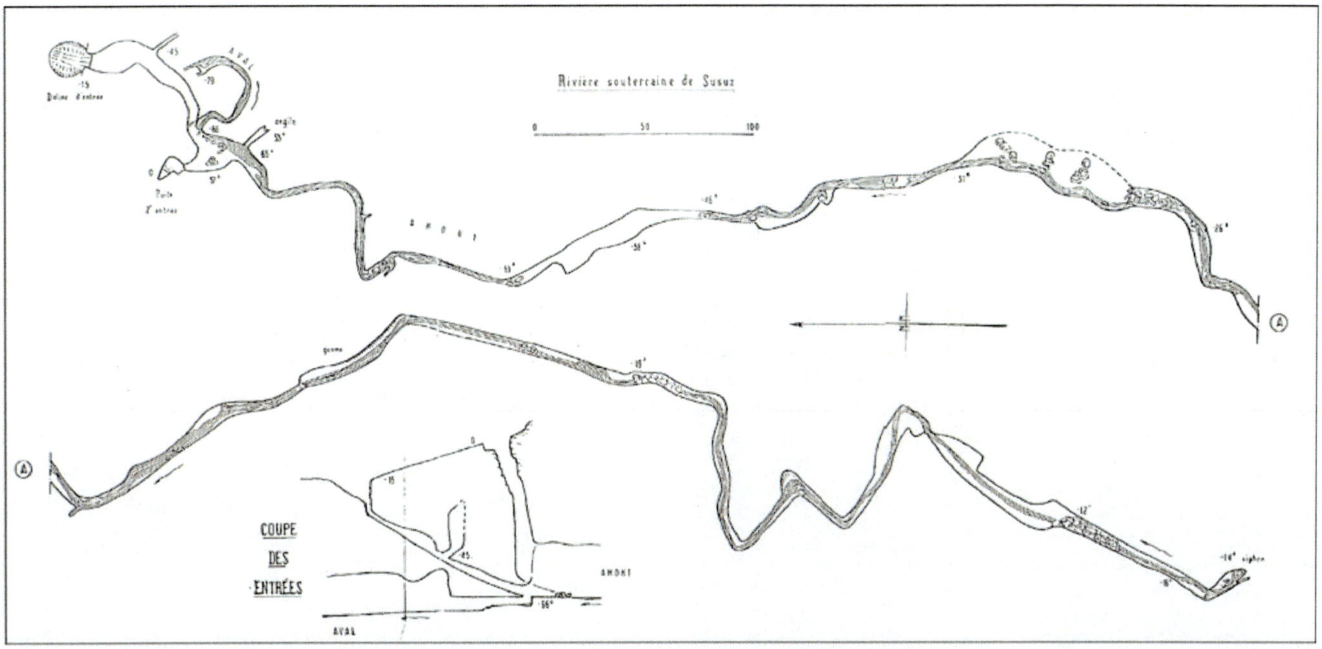

Fig. 4.53 Plan of Susuz Cave (Raimond & Chabert, 1971)

the main gallery that continues in the source-ward direction of the underground river. The ceiling height in these chambers, where piled rocks cover the river completely at some parts, reaches 15–20 m at some points.

The rest of the ground is flat, and the river is surrounded by sandbanks and gravels. Some parts of the main gallery are adorned with travertines, stalactites, stalagmites, and columns. Helictites were detected on some parts of the walls. The cave is branching into two galleries after 900 m. The western branch is 120 m long and ends with a sump. Piles of sandy gravel reaching up to 3 m high indicate that water mostly enters the cave through here. The sump is 125 m long and 16 m deep, only slightly changing its direction. It continues further in a spacious dry passage following the siphon. It widens at two points, creating high chambers. The ground is covered with gravel mainly and clay at some points. The passage is 380 m long and ends at another sump. The siphon at the end of the main gallery is only a few meters long, and it is followed by another one. The water amount coming into the cave from this point is far less than the waters coming from the siphons in the western gallery (Sroubek, 1989).

The earliest explorations and first survey of the Susuz Cave were conducted by Bakalowicz and Raimond (Bakalowicz, 1970; Raimond & Chabert, 1971) . The siphon divings and up-to-date survey of the cave were conducted by Geospeleos Team in 1989 (Sroubek, 1989).

4.2.39 Tilkiler Cave

Manavgat–Antalya
Total Length: 6818 m
Total Depth: −66/ +93 m

Tilkiler Cave is located 4 km northwest of Oymapınar Village of Manavgat District of Antalya, Southern Turkey. Manavgat District is one of the well-known areas in the Taurus Mountains with its complex hydrogeological potential harboring over 60 caves. The district is recognized by cavities that comprise the part of a hydrogeological system lying beneath the surface, as well as plenty of ravines and rivers. The fifth-largest dam in Turkey, Oymapınar Dam, was decided to be built here, and the Tilkiler Cave was revealed in 1974 when a natural gallery, with a powerful wind flow, was found in one of the injection tunnels during the explorations carried out for the construction of this dam. The injection tunnel that reaches to Tilkiler Cave after 450 m is 4 km west of the Oymapınar dam, at an altitude of 128 m, and at the contact point of conglomerate and impermeable rocks.

The cave is a part of the Eynif hydrologic system and functions as an overflow route for the Oymapınar resurgences (alt. 32 m, 10 m^3/s in winter). Dye tests that were carried out during the construction of the dam showed that the underground water of Tilkiler Cave is coming from Akpınar Cave, 35 km north (Aygen, 1984). The difference

Fig. 4.54 Tilkiler Cave First Lake (*Photo* A. E. Keskin)

of water levels between dry and wet seasons exceeds 60 m in the cave, and it was observed that there was a constant water flow of 12 m^3/s through the entry tunnel in spring. The cave is formed in Miocene conglomerates, which is a unit of Tefekli Formation, except for a short gallery formed within Burdigalian limestone at the southern extremity. Being formed mainly in conglomerates increases the value of the cave by rendering this cave to be one of the longest conglomerate caves in the world (Fig. 4.54).

The cavers of SCP and CMN first explored the cave in August 1976 and during the first year's survey, a total of 2755 m were mapped (C. Chabert, 1976). In 1977, the length reached 4845 m and the "muddy passage" was opened up. It carries a strong air current and was explored in 1978 (length 5585 m) and 1979 (J. Chabert, 1979b). Thirty years after the opening of the entrance, during the fifth expedition, it was observed that some galleries formed in the conglomerate have changed by corrosion even in this short period. Subsequently, in 2008, another new gallery was explored and surveyed. The total length of the Tilkiler Cave has reached 6818 m with the last exploration (Fig. 4.55). The main passage is named as the North Passage. Limestone overlies the conglomerate near the boulder slope leading to the Castle Room, and some scallop formations can be observed all the way along. The zone with clay dunes and the part that ends with a perfectly flat wall are the most remarkable areas of this passage. The large clay dunes in the North Passage beyond the Castle Room suggest that this passage is quite old. The flat wall owes its shape to the bedding plane between the conglomerate strata and the bed with smaller grains.

The "Upstream" part of the cave is a 425 m passage with a general direction of NNE. The main passage branches out the beginning of the "Upstream" part, which is a narrow downward muddy path on the right among blocks. This small crawlway opens up to a much smaller muddy passage. A slow but cold wind leads the air circulation in this passage that comes before a small pond followed by a 16 m pit. Following the muddy bottom of the pit, there is a small crawlway followed by a 30 m mud wall and four other crawlways that could only be passed by climbing. A continuous faint noise, which belongs to an underground spring, is heard in one of the highest spots (C. Chabert et al., 1978).

4.2.40 Tınaztepe Caves and Sinkhole

Seydişehir–Konya
Lengths: 1015 m and 1650 m (Sinkhole)
Depths: −5 m and −200 m (Sinkhole)

Tınaztepe caves and sinkhole reside 25 km south of Seydişehir on the side of the main road (Fig. 4.56). The system is composed of three fossil caves and an active sinkhole. The area had been through a complex geologic and geomorphologic transformation. When considering the area as a whole, young and old formations are observed to be intertwined. Tınaz Mountain is composed of nummulite limestones that comprise the top level of the comprehensive sequence. Gidengelmez Mountain found on the opposite side of this mountain is composed of Cretaceous aged sequences. The lengthwise depression in the middle, which is limited by two main fault lines, is composed of Paleozoic aged coat layers. Paleozoic aged limestones within ophiolitic enclaves are found next to flysh in this area. An allochthonous formation elongates as a thin belt in the northwest–southeast direction in front of the Tınaztepe Cave. Lower Paleozoic and Mesozoic aged limestones of this assembly come over the Eocene aged flysh formations of the Geyik Dağı Assembly. These two assemblies are separated by a large fault line that passes just in front of the Tınaztepe Cave System. The northwest–southeast direction fault line was effective on the large sizes of nummulite limestones in the area. Cretaceous limestones that compose the structure of

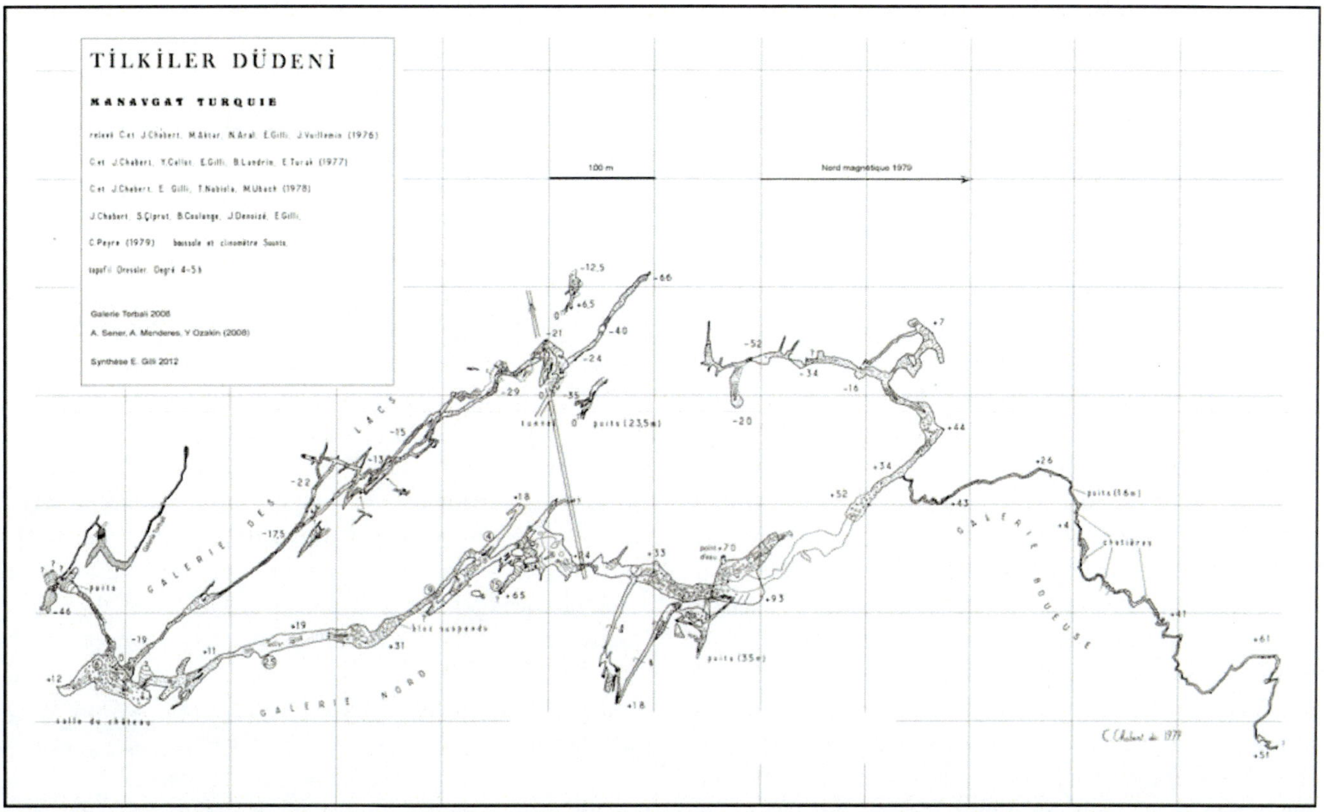

Fig. 4.55 Unpublished plan of Tilkiler Cave drawn by E. Gilli in 2012 upon the plan of (J. Chabert, 1979b) with an addition of a new passage explored in 2008

Fig. 4.56 Tınaztepe Caves, entrances of the first and second caves (*Photo* A. Yamaç)

Tınaztepe elongate toward the southeast and disappear under Paleozoic aged formations that composes the structure of Karadağ. Cretaceous limestones have a solid structure and their thickness changes between 350–1000 m. Limestones fragmented by cascaded fault lines are quite cracked. Gray or bluish-gray-colored Cretaceous limestones are found in a state of massive banks. Eocene flysh is seen as a northwest–southeast direction flysh corridor on the western side of the Tınaztepe Cave. The flysh sequence here is composed of limestone, sandstone, and marn alternations. Faultings developed according to neotectonic movements exacerbated at the end of the Miocene, affected the morphology of the mountain on a large scale. Eastern slopes of the mountainous area, where Tınaztepe resides, are fragmented by fault lines and gained a cascaded appearance. Declivity values reach 40–50° at some parts on these slopes (Bozyiğit & Meydan, 2017). Fault lines cutting Tınaztepe in various directions were effective in the creation and development of karstic formations. Tectonic properties are an effective factor on the course and development of karstification. Fault lines developed in various directions in the Neotectonic period caused the acceleration of karstification. The number of morphologic periods the discharge area in a closed basin state has been through can be observed from the cascaded plains and fossil caves found under and over the Tınaztepe Cave. The youngest system, which is yet under development stage and which drains the water of the area, is on the fault line 50 m under the fossil cave. The base level declined gradually during the water drainage of this closed basin, and this decline can be observed both inside the cave and in front of the cave in forms of terrace sets. Natural bridges formed in places where the old base did not collapse.

Tınaztepe Cave System developed on three different levels. The top-level cave developed at 1533 m while mid-level and bottom-level caves developed at 1500 m and 1440 m, respectively. Tınaztepe Cave System elongates toward the Suğla Lake to the east. Today, the waters sinking at the entrance of this cave resurface near the Susuz Village. The cave formed at the uppermost fraction of the Tınaztepe Cave divides into two parts starting from its entrance, and each part is about 100 m long. These caves are completely dry. Stalactite and stalagmite formations are observed within. The cave developed in the middle level is composed of three galleries; one large and two small ones. The large one called the Big Tınaztepe Cave is 1015 m long (Güldalı, Onal, & Nazik, 1982). The roof height reaches up to 30–40 m in the middle and final sections, following a low but wide entrance. Wall travertines, stalactites, and stalagmites can be seen in the faults and fractures crosscutting the cave. Small streams and ponds are common during the high season. It becomes dry except the pond in the final section during other seasons. The two small branches next to the main gallery, the branches A and B, are approximately 100 m long each, and cave B is completely dry. Stalactites, stalagmites, and columns were formed at places where faults crosscut the cave. The cave is connected with the other fossil cave through a chimney due to the subsequent collapse in the middle section.

The cave that functions as a doline at the bottom level is called the Tınaztepe Sinkhole. A stream that originates around Ağaçtepe and reaches a flow rate of 1 m^3/s from time to time ends at this sinkhole. Due to Tınaztepe's ascension through young tectonic movements, this stream continuously eroded its bed deeper, developed the Tınaztepe Cave System by forming the upper- and lower-level caves in this order and reached to the level it flowed through today. Waters pour into this doline at the lowest level, from underneath of a naturally formed bridge. This natural bridge shows that the previous bed of the stream was higher. The connection with the Susuz Cave, which has an altitude of 1172 m and a distance of 7 km, has been proven by dye tracing (Güldalı et al., 1982). The cave ends with a siphon in the final zone. Transportation of the diving equipment is troublesome because of the pond in the main corridor and numerous small descents. Only 18 m were advanced because the siphon continues along with a very narrow gallery. Tinaztepe Cave System was first explored and mapped by the cavers from SCP in 1968 (Fig. 4.57) (Bakalowicz, 1968) followed by MTA in 1980 (Güldalı & Nazik, 1984; Güldalı, Nazik, & Onal, 1980).

4.2.41 Üçağızlı Cave

Samandağ–Hatay
Total Length: 30 m

Üçağızlı Cave is 12 km south of Samandağ county of Hatay, on the shore of the Mediterranean Sea (Fig. 4.58). The cave is on the slope of Kel Mountains, by the seaside with an altitude of 18 m. The southwest-facing entrance of this 30 m deep, 200 m^2 cave that formed within Cretaceous calcareous rocks, is opened up due to collapses. It is known that numerous caves of various sizes are found within the same geological formation in the area. Üçağızlı Cave was discovered and excavated for the first time by A. Minzoni-Deroche in 1989–1991(Güleç et al., 2003). Later in 1996–1997 E. Güleç has started a research project in Samandağı, and a test excavation was carried out in 1997 followed by a second excavation in 1998 which is still ongoing. Üçağızlı Cave is an Upper Paleolithic settlement dated between 27,000 and 42,000 BP and had been continuously inhabited by humans for 12,000 years. The Ahmarian layer constitutes the final layer of culture in the Upper Paleolithic cave. Being different from Aurignacian culture, this culture is characterized by blade/bladelet tools at the upper layers, unretouched tools at the lower layers, and the

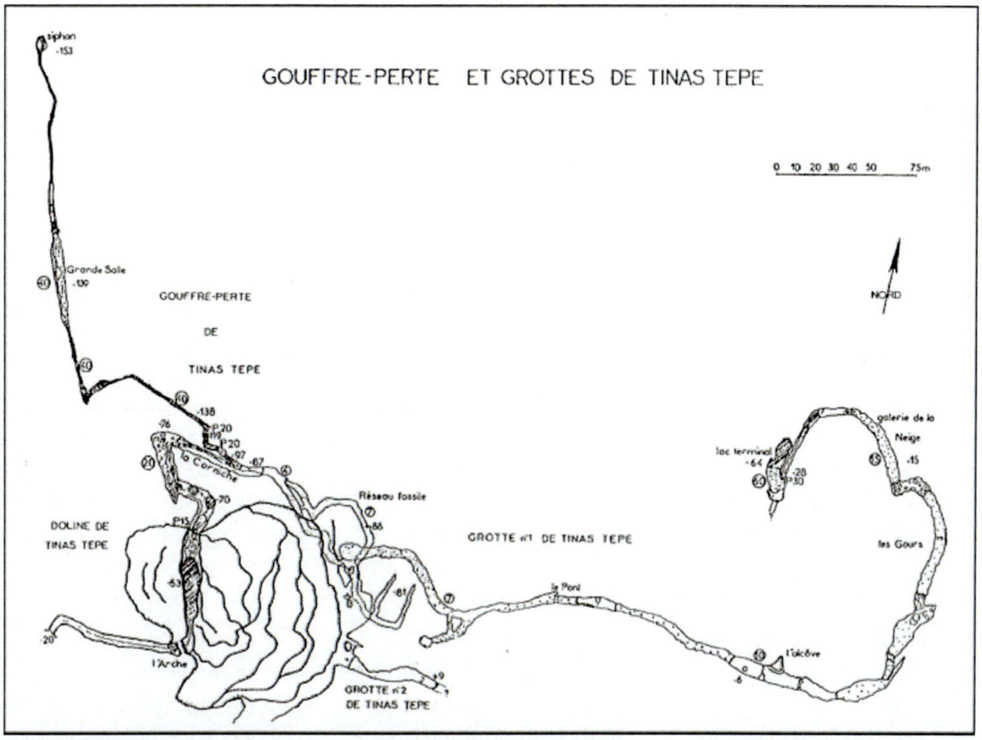

Fig. 4.57 Plan of Tınaztepe Cave System (Bakalowicz, 1968)

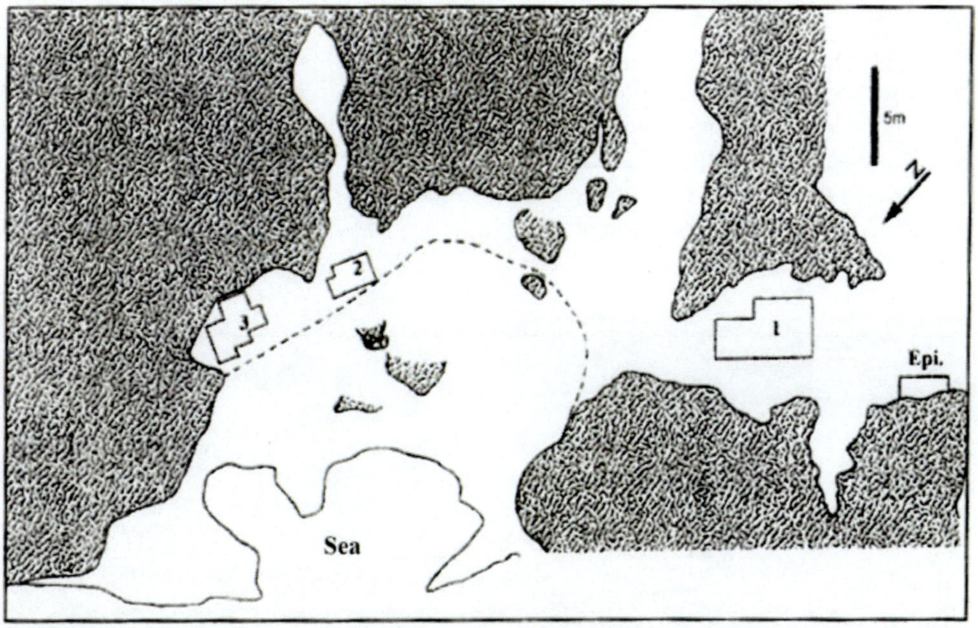

Fig. 4.58 Plan of Üçağızlı Cave adapted from (Güleç, Kuhn, Ozer, & Steiner, 2003)

presence of the proto-prismatic cores. The features of Ahmarian cultures in Üçağızlı Cave show that their Upper Paleolithic culture is within the Levantine. The most important paleoanthropological finds of Üçağızlı Cave are tooth and skeleton remains of human beings that lived in the cave during the Paleolithic Period. Fourteen different human teeth were found during the excavations in addition to maxilla and skull fragments. They were dated between 29,000 and 41,000 BP. Morphologies of deciduous and adult teeth found within Üçağızlı Cave show characteristics of modern humans. Measurements of the teeth are bigger than both of late Pleistocene people's and modern humans. Thus, inhabitants of the Üçağızlı Cave were Homo sapiens that have some archaic characteristics (Güleç et al., 2009, 2010). There are bones of animals among the finds, which were used both for nutrition and adornment, that suggest a very rich fauna; such as shell remains of marine and land mollusks, bones of bezoar goat, fallow deer, red deer, roe deer, gaur, grizzly bear, weasel, wildcat, birds, and fish of different sizes, hare, and tortoise (Güleç et al., 2003). A huge number of stone tool samples and a few bone artifacts were also found. The extensive ornament assemblage consists of small beads or pendants made from modified marine shells. A variety of Mediterranean species were used in the cave, and more than a thousand shell ornaments were recovered. The pierced and painted shells found in layer H are among the earliest indisputable personal ornaments in the world. They are dated between 39,000 and 41,000 BP (Güleç et al., 2003).

4.2.42 Yarımburgaz Cave

Küçük Çekmece–Istanbul
Total Length: 1021 m

Yarımburgaz Cave lies approximately 22 km southwest of Istanbul in the district of Küçük Çekmece, about 1.5 km north of the northern shore of the Küçük Çekmece Lake (Fig. 4.59). Opened through abrasion of the underground waters on the western slope, it resides on an Eocene origin calcareous formation; it consists of two interconnected sections with diverse forms and dimensions as well as diverse entrances at different levels. It is around 11–18 m high from the Marmara Sea. There are two sections roughly elongating in the northeast–southeast direction. The upper section is like a large hall with a dimension of about 15 × 52 m and a present height of 10 m. It was probably used as a church during the Byzantine period after some adjustments. The holes on the walls for bond courses indicate that there used to be a setback story with a tiled roof. This church is certainly related to the monastery building outside of the cave. The lower section, which initiates like a gallery and continues as a tunnel toward the depths of the rocks meandering and composing some large halls, is accessed from the upper level through a ramp near the entrance. Although the upper cave is occluded immediately, the lower one continues about 600 m. Approximately 240 m after the entrance in the lower cave following the bifurcation, the right branch elongates a little more than the left branch and becomes impassably narrow preventing any further progress after a somewhat wide vestibule at the end. The formation marks of the cave can be traced by the presence on a steep valley eastward of Sazlidere. This sedimented valley was likely an elongation of the lower cave during the Lower Paleolithic Age. The bedrock was reached only at the entrance in the lower cave during the excavations, and the actual height of the cave was not determined since it could not have been descended. It is estimated that the environmental conditions during the Paleolithic Age were very different from the present conditions.

Yarımburgaz Cave is subject to countless speleological and archaeological research due to its closeness to İstanbul. The first known cave research in Turkey was conducted by Abdullah Bey in 1869 in Yarımburgaz. It was followed by another article of the same author in 1872 (Bey, 1872), a field report of R. Bousquet published in Echos d'Orient magazine in 1901 (Bousquet, 1901), a conference text of H. Kocacan published in Tedrisat Mecmuası in 1918 (Kocacan, 1918), and a quite extensive study and survey conducted by Professor Hovasse in 1927 (Hovasse, 1927). There are five different maps of Yarımburgaz Cave drawn on different dates (Fig. 4.60).

Fig. 4.59 Entrance of Yarımburgaz Cave (*Photo* A. Yamaç)

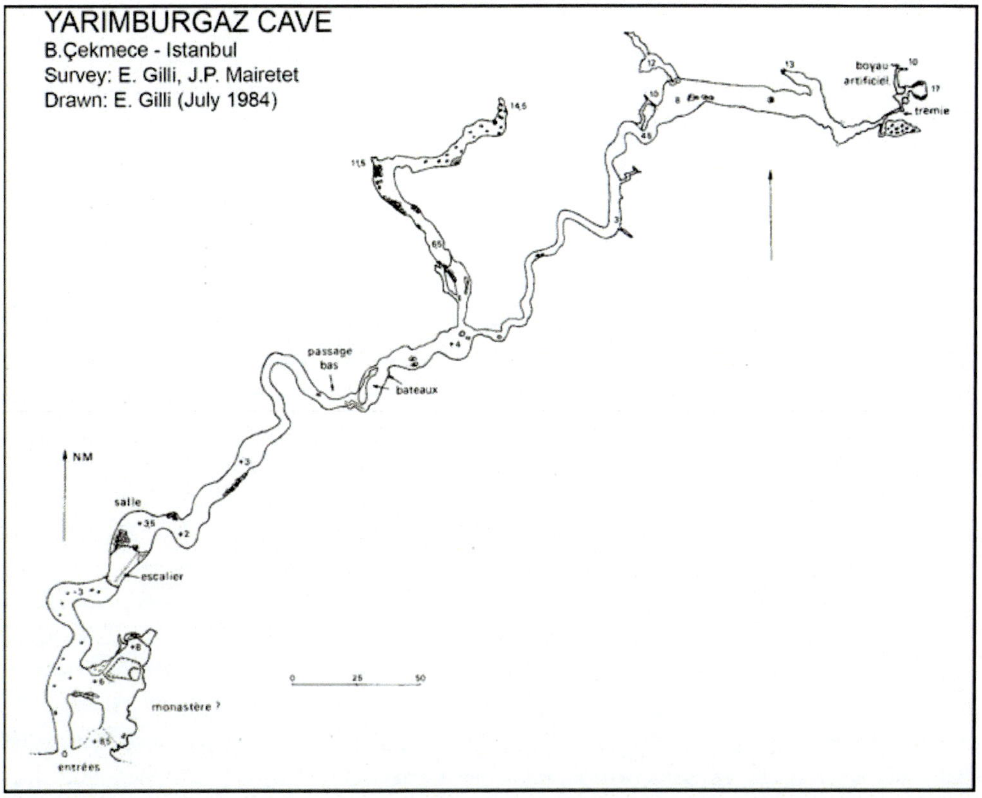

Fig. 4.60 Unpublished map of Yarımburgaz Cave drawn by E. Gilli

The first excavations carried out at the site under the leadership of Ş.A. Kansu and K. Kökten in the entrance of the lower cave in 1963 (Kökten, 1963) were later continued more systematically under the direction of Kansu, Kökten, and N. Dolunay in 1964 and 1965 (Kansu, 1966). The site suffered damage in the twenty years between 1966 and 1986, and this instigated the second round of excavations which were carried out in 1986 by M. Özdoğan. This salvage excavation revealed that the oldest occupation in the cave dates to the Lower Paleolithic, and the third round of excavations was undertaken to further understand Turkey's Pleistocene. They were conducted jointly by G. Arsebük and F. Clark Howell of the University of California at Berkeley between the years of 1988 and 1990.

In the first phase of excavations carried out in Gallery A of the lower cave, Byzantine sherds within the 40 cm top layer and the 60 cm second layer with pebble tools and debitage within the third yellowish silty and sandy layer were unearthed. This stratum most probably dates to the Middle Paleolithic according to Kökten. Bedrock was reached 370 cm below the surface. Kansu and Kökten report that teeth and bones of Ursus spelaeus cave bear were found in association with coarsely made points in this third stratum. It was revealed in the excavations in the upper cave referred to as "Gallery B" that it has the same stratum with Gallery A and beneath these, there is a 200-cm-thick Chalcolithic deposit stratum (Kansu, 1966).

The second phase of excavations in 1986 was more detailed and systematic. A very different stratigraphy emerged in this excavation directed by Özdoğan. Upper cave strata 15-12 and lower cave strata 12-3 were dated to the Lower Paleolithic. Findings of pebble tools suggested that these Yarımburgaz strata belonged to the beginning of the Middle Pleistocene and earlier periods (Özdoğan, 1988). This was proven wrong in the third phase of excavations led by Arsebük. The second phase of excavations dated Yarımburgaz Cave's earliest levels to one million BP. However, further research has made it clear that they are slightly more recent. The second-lowest deposit dated to the Middle Pleistocene by Özdoğan includes strata 11-8. Habitation in this period is believed to have been intermittent. Immediately above this deposit, Özdoğan identified Upper Paleolithic and Epipaleolithic deposits in the upper caves in the 7th and 6th strata. The 5th stratum in the upper cave represents the Neolithic era. Again in the upper cave, strata 4-2 belongs to the Chalcolithic following the Neolithic (Arsebük, Howell, & Özbaşaran, 1992). Findings from this period were found within a natural depression in the interior of the upper cave, and the first stratum is the only one that is the same in both caves. Stratum 1 included a dressed-stone wall built at the point where the two caves meet. This stone structure is probably the apse of a church dating to the earliest Christian times. The excavations led by Arsebük were aimed primarily at shedding light on the Pleistocene Epoch. Geomorphological and paleo-ecological studies have accompanied the archaeological excavations in and around the cave area. Arsebük has divided the Pleistocene deposits into four different geological categories. The first two of these are conglomerate layers are composed of quartz rubbles. The surfaces of these layers have cement-like appearances. There is a silty dark-reddish brown layer underneath. This third stratum differs from the first two strata by being comprised of genuine Pleistocene deposits, and it is rich with material culture (Arsebük, Howell, & Özbaşaran, 1990).

The first occupants of Yarımburgaz made use of flint cores, chert, quartz, nodules of various sizes, and other raw materials, chipping them to produce tools for everyday use. The majority of the findings belonging to the Yarımburgaz chipped stone industry are flake types. 80% of the chipped stone artifacts found consist of retouched flakes with steep retouches, simple flakes, and other flake-type tools. Arsebük believes that the Yarımburgaz flint-work techniques resemble the Near Eastern Tayac industry rather than their European counterparts, and he further argues that the people living in the second half of the Middle Pleistocene in Yarımburgaz used functional tools that lacked unnecessary details. He dates this cultural deposit to 450,000–130,000 BP (Arsebük, Howell, & Özbaşaran, 1991). Excavators screened microfaunal remains in the third phase of excavations to document the paleo-ecology of the Pleistocene and the pre-Pleistocene. Remains of bats, birds, various rodents, and fish were found. The fish bones are especially interesting. It is not certain whether they were brought into the caves by humans. The fossilized bones of an extinct variety of the modern mountain bear and Ursus Deningeri are also of particular importance. Many of the bones of this mammal were found in association with chipped stone tools.

The electron spin resonance (ESR) method applied on ten Ursus Deningeri teeth to detect the archaeometry dating of this layer, which Yarımburgaz Lower Paleolithic Age findings are gathered from, gave the date results of 400,000 BP (Bulur, Ozer, Goksu, Cetin, & Unal, 1991). Arsebük believes that the cave was used alternately by bears and people and supposes that human occupation was limited to spring, summer, and fall. There were jackal or fox gnaw marks on the bear bones but there were no stone tool marks that indicate human cutting or scraping. Bones of other herbivorous and carnivorous animals including animals from the canine, feline, equine, hyena, deer families, and some horned animals have also been found.

4.2.43 Yaylacık–İnilti Pazarı Cave System

Gündoğmuş–Antalya
Total Length: 5929 m
Total Depth: −595 m

The first exploration of Yaylacık–İnilti Pazari Cave System, which is 15 km east of Gündoğmuş county of Antalya at an altitude of 2160 m, started in 2011. It was understood that there were two different but connected sinkholes feeding an underground river, as a result of the research conducted by MAD. The area is composed of units that reside in the Geyik Mountain Assembly. Geyik Mountain, in the Mid-Taurus region, one of the highest mountains in the region, gave its name to the formation. Different formations with tectonic contacts, that reflect stratigraphically and structurally different ambient conditions, are known to be situated along the Mid-Taurus belt. These formations, each carrying different tectonostratigraphic properties, sustained displacements on a scale of kilometers due to movements in Lower Cretaceous and Mid-Eocene. Tectonic lines within Geyik Mountain Assembly in the Mid-Taurus region roughly elongate in the northwest–southeast direction (Baykara, 2018). The tectonically ascended area became suitable for the development of cave systems via streams sinking underground through sinkholes. Yaylacık–Inilti Pazarı Cave System is essentially a semi-active cave system formed within Permian–Carboniferous limestones and the Lower Cretaceous limestones incompatibly residing over it. Many fault lines are elonging in the southeast–northwest direction in the area (Şimşek, 2018).

Yaylacık Sinkhole continues with a 90 m descent following a short horizontal passage and presents a vertical development in contrast to all the caves researched in the Mid-Taurus region. The cave that consistently continues with tight meander galleries following the 90 m descent mostly has short descents and a constant declivity. This formation changes completely at -319 m and the cave that consists of narrow galleries up to this point, suddenly reaches an underground river flowing in a very high and wide gallery. The entrance of one of the branches reaching this underground river, which feeds on numerous side branches, is the Inilti Pazarı Cave that resides very close to the entrance of Yaylacık Sinkhole. This second sinkhole joins the Yaylacık Sinkhole at the depth of −327.

Yaylacık–Inilti Pazarı Cave System is important as it is the first horizontal cave system found at high altitudes in Mid-Taurus Range and the first known example of a sinkhole reaching an underground river (Fig. 4.61). Even though the Yellibel Cave and the Çadırçukur Cave, which are in the same region and very close to these two caves, end at −229 m and −169 m respectively, the total length of this large underground cave system is expected to increase with the discovery of new connections in the future.

Fig. 4.61 Yaylacık Cave (*Photo* Y. Özakın)

4.2.44 Yazören Cave

Savaştepe–Balıkesir
Total Length: 3564 m
Total Depth: −113 m

Close to Savaştepe Village of Balıkesir, Yazören Cave is one of the longest traverse caves of Turkey. Yazören Cave is a semi-active sinkhole with a sinking creek and a resurgence mouth. The geological structure of the area is quite complex. Limestones that outcrop in the area are generally dated to Mesozoic Era in Y. Hoşgören's article (Hoşgören, 1981). It is shown to be dated to Triassic and Jurrasic periods in two different maps. There are Late Miocene–Early Pliocene dated volcanic tuff, marn, sandstone, and andesite residing on the limestones, which has an appearance of a plateau at an approximate altitude of 700–800 m, discordantly. This whole construct gained a quite complex form due to the live tectonic movements in the area. The 12 faults in just 5 km distance between two villages are proof of the tectonic activity in the area. Hoşgören remarks in his article that the limestone formation found on the plateau has a solid structure with a thickness of 500–600 m. There are mainly average-sized many dolines and uvalas on the plateau toward the south, east of Bozarmut and Sultaniye villages. Some of these dolines and uvalas become lakes during the rainy seasons. The most interesting formation in the area other than these is the polje known as the Yazören Plateau. Probably being the largest polje in Northwest Anatolia, this depression is approximately 2.5 km long and 1 km wide (Hoşgören, 1981).

Fig. 4.62 Yazören Cave (*Photo* Ç. Çankırılı)

Following the Ayvaini Cave, Yazören Cave is the second-longest traverse cave explored in Turkey, and it starts with a large entrance (Fig. 4.62). Aside from the main gallery of the cave, there is a serious amount of water input coming from the side branches. There are signs of ponding, up to the point of becoming a siphon in the rainy season, at some points of the main gallery. There are large amounts of green and gray andesites among the rocks that compose the ground of the main gallery. As written in Hoşgören's article, it indicates that this material mismatching the structure of the cave was carried into the cave by waters from the andesite area residing north of the doline. The amount and size of these rocks carried through the main gallery and the gravels stuck into cracks merging show that the flow rate in the cave is high during the rainy season. There are no speleothems observed starting from the entrance point and along the entire cave, except a fossilized branch almost 500 m after the entrance. Aside from conglomerate and breccia discordances, there is an intense amount of stalactites and stalagmites inside this westward fossil branch. The ceiling height of the main gallery reaches 15 m at some parts, and it becomes a siphon after a short distance. The gallery becomes high and wide again following this part, which is 45 cm high and filled with mud in the dry season. The gallery found in the middle of the main gallery is considered to be a side branch, despite continuing like the main gallery. It is probable that the resurgence of this gallery collapsed and blocked the gallery, causing the water to carve out a new way through another crack toward the west.

Although no living animals were encountered in the active gallery, the entrance and the fossil gallery contain a large number of bats. A huge amount of bat population inhabit the side branches. Rhinolophus hipposideros and Myotis myotis were observed. Besides bats, there are many other living creatures in the side branches. Spider types stick out among these. Yazören Cave was explored and surveyed by ASPEG in 2008 (Yamaç, 2009) (Fig. 4.63).

4.2.45 Zindan Cave

Aksu–Isparta
Total Length: 760 m
Total Depth: +15 m

Zindan Cave is located in Western Taurides, about 2 km northeast of Aksu county of Isparta, next to the Aksu creek. The thickest and most common limestones in the region are Cretaceous and Jura aged. Antalya Nappe, which is composed of highly complex rock units found in an allochthonous state, resides west of this autochthonous belt that spreads to a very wide area. Zindan Cave developed within Jura–Cretaceous aged breccia and turbiditic limestone formations of this nappe. The thickness of this formation is about 100–300 m. These limestones are quite pure and faulted. The ceiling height of the cave reaches 15–20 m at parts that coincide with faults. Zindan Cave generally developed on faults that elongate in a north–south direction. The entrance of the cave looks like an ellipse with a diameter of 6×2 m , and it is 12 m high from the Aksu creek. The cave turns toward the north with two turns just after the entrance. The ceiling height of this part is 20 m, and there is a small bat colony within. Although the cave narrows and lowers down between 50 and 105 m, width and ceiling height increase again following the 105 m. Large cauldrons and a dense gravel deposit are observed following the 200 m. These show that a strong underground creek used to flow here. There is a small side siphon at the 355 m, and there are many stalactites and stalagmites at this part of the cave. Sinkholes are encountered between 425 and 450 m, where waters disappear by leaking underground. The average width of the gallery is 2 m between this point and the part at the 640 m. Walls are covered with stalactites and stalagmites while the ground is filled with gravel and sand in this section (Fig. 4.64). A small creek with a depth of 5–6 cm flows toward the exit here. Ceiling height lowers down, and the underground creek gets deeper following this section. The ceiling lowers down and creates a siphon at 760 m, making it impassable. Considering the profile of the Zindan Cave, its horizontal elongation with a mild declivity in such a long distance reminds that the floor formed on a rock formation that does not have karstification properties, but such type of formation is not encountered on the floor of any part of the cave. On the other hand, sizes of sand piles and pebbles inside the cave and the depth and width of the cauldrons show that there was a large underground creek

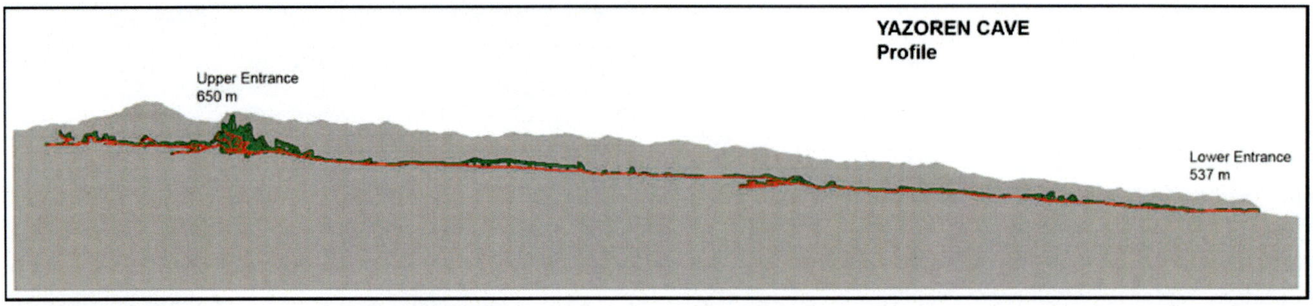

Fig. 4.63 Profile of Yazören Cave (Yamaç, 2009)

Fig. 4.64 Zindan Cave (*Photo* M. Kızıl)

here in geological periods past. There is just a weak creek flowing inside the cave today (Güldalı, Nazik, Soylu, Aksoy, & Beydeş, 1989).

A Eurymedon statue was found during a tunnel construction close to the entrance of the Zindan Cave in 1977, and the cave was defined as Eurymedon Sanctuary. A Eurymedon head and a mosaic with dolphin patterns on its sides were found inside a panel with black, white, and red stones on it, during the excavations conducted at the entrance of the cave between 2002 and 2004. Among the architectural tilings inside the cave, only holes for stones found on the rocks were preserved. The three terraces in front of the cave were made of ancient blocks. As a result of some studies, it was understood that the cave was reached through a road with stone pavements that could be dated to a late ancient age. Aside from the road with the stone pavement, a 200 m second tiling made of pink rocks that belong to an early period was found in the middle part of the terrace, and this composition was joined with the main rock. Holes for stones and niches of religious purposes are found at the rocky area east of the terrace. There is a podium with two preserved lines which is composed of bossaged blocks protected in situ in front of the cave and floor tilings in front of the podium. The podium and the area are re-arranged with the blocks of the floor tiling and the inscriptions found on the podium dated back to AD 3–4 century. The three-door lintels found between architectural blocks and memorial pieces that could shed light on the architectural arrangement of the sanctuary are in their original sizes, and their difference in profiles indicates that there are three different entrances in the area. Aside from the size differences of garlanded friezes, the scale differences of architrave blocks also indicate that there is a rich architectural structuring in this religious area. Sacrificial stele attributed to Zeus and Kybele is also important for this place to be defined as a sanctuary (Öcal, 2012).

Even though a rich bat colony was observed during the research conducted by Benda and Horacek, this bat colony became smaller after the cave opened to tourism. Myotis myotis, Miniopterus schreibersii, a limited number of Myotis cappacini and Myotis oxygnatus were found among the species detected (Benda & Horacek, 1998).

The first research of Zindan Cave was conducted by Coffait in 1959 (Spitzenberger, 1973). It was re-explored and surveyed M. Bakalowicz in 1968 (Bakalowicz, 1968). It was explored and mapped again by MTA and Sheffield University Speleological Society separately in 1989 (Fig. 4.65) (Gregory, 1989; Güldalı et al., 1989).

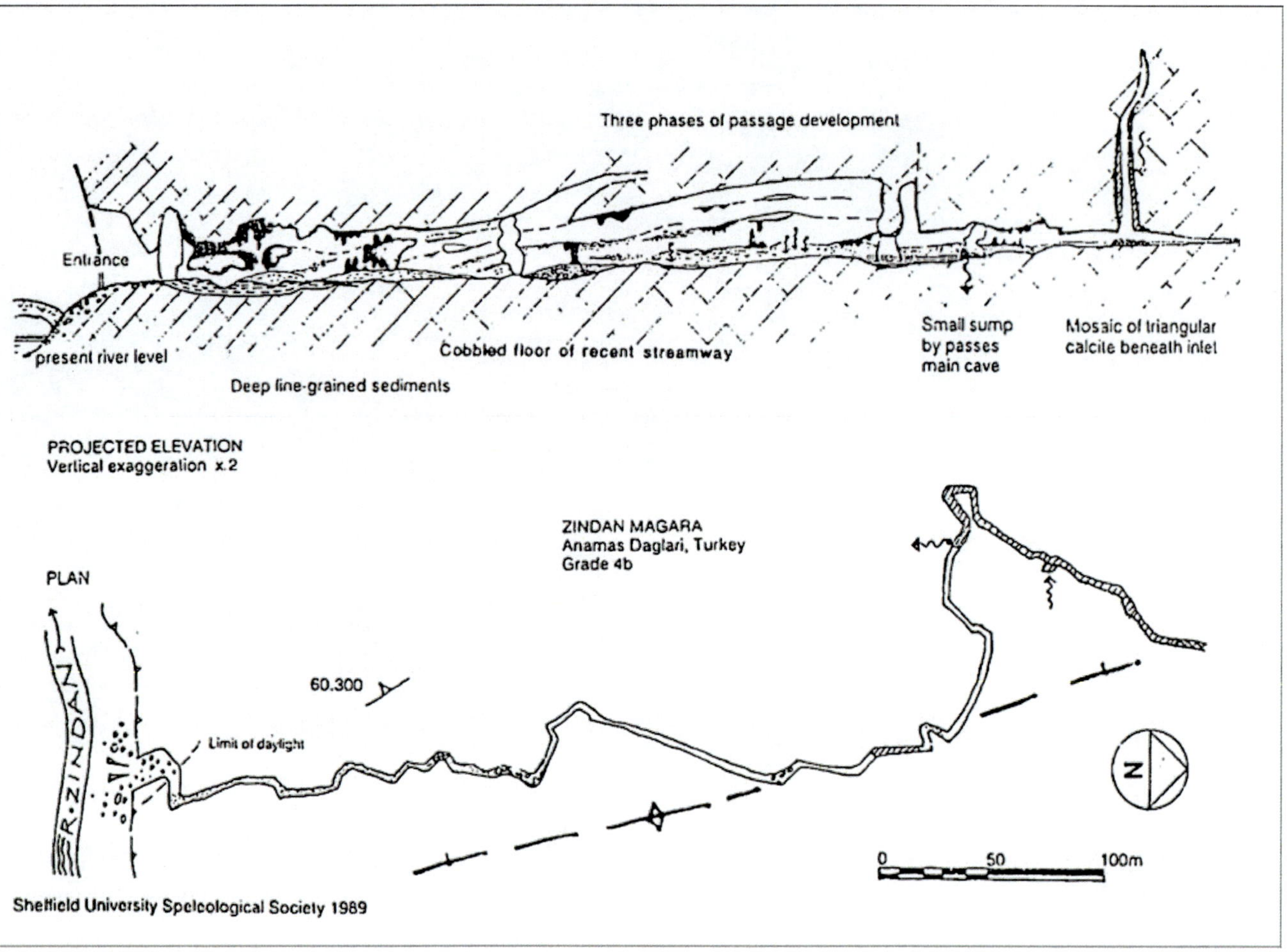

Fig. 4.65 Profile and plan of Zindan Cave (Gregory, 1989)

References

Akay, E., & Uysal, S. (1988). Orta Toroslar'ın Post-Eosen Tektoniği. *MTA DERGİSİ, 108,* 57–68.

Akkemik, Ü., Aytuğ, B., & Güzel, S. (2004). Archaeobotanical and Dendroarchaeological studies in İlgarini Cave (Pınarbaşı, Kastamonu, Turkey). *Turkish Journal of Agriculture and Forestry, 28*(1), 9–17.

Alagöz, C. (1944). *Türkiye Karst Olayları Hakkında Bir Araştırma (Une étude sur les phénomènes karstiques en Turquie).* Ankara.

Altiner, D., Bati, Z., Tunay, G., Senel, M., & Ekmekçi, E. (2006). *Trakya Bölgesi Litostratigrafi Birimleri.* Ankara: MTA.

Arsebük, G., Howell, F., & Özbaşaran, M. (1990). *Yarımburgaz 1988.* Paper presented at the 11th Kazı Sonuçları Toplantısı, Ankara.

Arsebük, G., Howell, F., & Özbaşaran, M. (1991). *Yarımburgaz 1989.* Paper presented at the 12th Kazı Sonuçları Toplantısı, Ankara.

Arsebük, G., Howell, F., & Özbaşaran, M. (1992). *Yarımburgaz 1990.* Paper presented at the 13th Kazı Sonuçları Toplantısı 1991, Çanakkale

Atabey, E., Nazik, L., & Törk, K. (2001). *Oylat mağarası çökelleri ve depolanma mekanizması.* Paper presented at the 54. Türkiye Jeoloji Kurultayı, Ankara.

Atabey, E., Nazik, L., & Törk, K. (2002). Oylat Mağarası Çökel Kayalarının Sedimantolojisi, İnegöl (Bursa). *MTA DERGİSİ, 123–124,* 91–98.

Atalay, E., & Ülkümen, O. (1980). *Kırklareli-Dupnisa Mağara Sistemi Araştırma Raporu.* Retrieved from İstanbul.

Atalay, I., Karadogan, S., & Yıldırım, A. (2010). *Karstification and Ground River System in SE Anatolia: A key Study from Birkleyn Cave System.* Paper presented at the 7th Turkey-Romania Geographical Academic Seminar, Antalya.

Aygen, T. (1984). *Türkiye Mağaraları.* İstanbul.

Bakalowicz, M. (1967). Hydrogéologie du Taurus Calcaire Entre Beyşehir et Manavgat. *Grottes et Gouffres, Bulletin Périodique du Spéléo-Club de Paris, 40,* 29–37.

Bakalowicz, M. (1968). Données Géologiques, Hydrologiques et Météorologiques sur les Cavités, Pertes et Émergences Reconnues Pendant la Campagne 1968. *Grottes et Gouffres, Bulletin Périodique du Spéléo-Club de Paris, 42,* 27–42.

Bakalowicz, M. (1970). Campagne Spéléologique 1970 en Taurus Occidental Expédition française en Turquie. *Grottes et Gouffres, Bulletin Périodique du Spéléo-Club de Paris, 45,* 15–27.

Bayarı, S., Ozyurt, N., & Pekkan, E. (2009). Giant collapse structures formed by hypogenic karstification: The obruks of the Central Anatolia, Turkey. In A. Klimchouk & D. Ford (Eds.), *Hypogene Speleogenesis and Karst Hydrogeology of Artesian Basins* (pp. 83–90). Simferopol: Ukraine Institute of Speleology and Karstology.

Bayarı, S., Ozyurt, N., Törk, K., Avcı, P., Güner, I., & Pekkan, E. (2017). Geodynamic control of Hypogene Karst development in Central Anatolia, Turkey. In A. Klimchouk, A. Palmer, J. De

Waele, A. Auler, & P. Audra (Eds.), *Hypogene Karst Regions and caves of the world* (pp. 449–462): Springer

Bayarı, S., Varinlioglu, G., Keskin, A., & Erdem, B. (2002). *Insuyu Mağarası'nda Güncel Sualtı ve Suüstü Bulguları*. Paper presented at the Sualtı Bilim ve Teknolojileri Toplantısı, İstanbul.

Bayarı, S., Zreda, M., Ciner, A., Nazik, L., Törk, K., Ozyurt, N., ... Sarıkaya, A. (2003). *The Extent of Pleistocene ice cap, glacial deposits and glaciokarst in the Aladaglar Massif: Central Taurids Range, Southern Turkey*. Paper presented at the XVI INQUA Congress, Nevada.

Baykara, M. (2018). Ümitli Yaylası'nın (Gündoğmuş, Antalya) Jeolojik ve Jeomorfolojik Özellikleri *Pamukkale Univ Muh Bilim Dergisi, 24*(3), 554–560.

Benda, P., & Horacek, I. (1998). Bats of the Eastern Mediterranean. Part 1. Review of distribution and taxonomy of bats in Turkey. *Acta Societatis Zoologicae Bohemicae, 62,* 255–313.

Bey, A. (1872). Études Géologiques sur les Environs de Constantinople, Yarym Bourgas, Macri-Keuy, Sari Keuy. *Gazette Médicale d'Orient, Constantinople, Octobre, 7,* 103–105.

Bousquet, R. (1901). Les Grottes de Yarim-Bourgas. *Échos d'Orient-Stamboul, 4,* 295–302.

Boztuğ, D. (1992). Daday - Devrekani Masifi Güneybatı Kesiminin Litostratigrafi Birimleri ve Tektoniği. *MTA DERGİSİ, 114,* 1–20.

Bozyiğit, R., & Meydan, A. (2017). Tınaztepe Mağarası ve Çevresinin Jeopark Potansiyelinin Değerlendirilmesi. *International Journal of Eurasia Social Sciences, 8*(29), 1131–1151.

Brignoli, P. (1971). Un nuov(e)o Troglohyphantes cavernicolo ed anoftalmo (ouanophthalmo) dell'Asia minore (Araneae, Linyphiidae). *Fragmenta Entomologica, 7/2,* 73–77.

Brignoli, P. (1972). Terzo contributo alla conoscenza dei Ragni cavernicoli di Turchia (Araneae). *Fragmenta Entomologica, 8*(3), 161–190.

Brignoli, P. (1973). *Zoogeographical observations of the caves Spiders of the MiddleEast*. Paper presented at the 6th International Congress of Speleology, Olomouc.

Bulur, E., Ozer, A., Goksu, H., Cetin, O., & Unal, G. (1991). *Yarımburgaz Mağarası Travertenleri ve Diş Minelerinde ESR Çalışmaları*. Paper presented at the 6th Arkeometri Sonuçları Toplantısı 1990, Ankara.

Chabert, C. (1966). Turquie 66. *Grottes et Gouffres, Bulletin Périodique du Spéléo-Club de Paris, 38,* 2–21.

Chabert, C. (1967). Lettre de Turquie. *Grottes et Gouffres, Bulletin Périodique du Spéléo-Club de Paris, 40,* 5–14.

Chabert, C. (1972). Chap.V - De l'expédition à Pınargözü en l'an 1971. *Grottes et Gouffres, Bulletin Périodique du Spéléo-Club de Paris, 48,* 5–28.

Chabert, C. (1975). Enigmes turques. Du massif du Dedegöl dağı aux gorges de la Manavgat (Taurus). *Grottes et Gouffres, Bulletin Périodique du Spéléo-Club de Paris, 55,* 3–11.

Chabert, C. (1976). *Recherches sur les Systèmes de Kembos et d'Eynif (Taurus, Turquie 1976)*. Paris.

Chabert, C. (1977). Pınargözü, Caverne du vent, La Montagne et Alpinisme. *Club Alpin Français* (4), 192–197.

Chabert, C., Callot, Y., Chabert, J., & Gilli, E. (1978). Les recherches de 1977 sur le système hydrologique d'Eynif (Taurus occidental, Turquie). *Grottes et Gouffres, Bulletin Périodique du Spéléo-Club de Paris, 67,* 3–28.

Chabert, J. (1979a). Kocain, la "grotte énorme" (province d' Antalya, Turquie). *Grottes et Gouffres, Bulletin Périodique du Spéléo-Club de Paris, 72,* 21–29.

Chabert, J. (1979b). La campagne 1978 en Turquie (Manavgat, Antalya). *Grottes et Gouffres, Bulletin Périodique du Spéléo-Club de Paris, 71,* 3–22.

Çetin, E., Süleymangil, H., Tarba, G., & Ulkümen, O. (1983). Kastamonu Bölgesi Mağara Araştırmaları. *DELTA, 1,* 3–12.

Değirmenci, M., Bayarı, S., Denizman, C., & Kurttas, T. (1994). Caves in Conglomerate, Köprüçay Basin (Western Taurids, Turkey). *Journal of Caves and Karst Studies, 56,* 14–22.

Döker, M. (2013). Çem Düdeni. *DELTA, 8,* 79–83.

Döker, M., Özakın, Y., & Güloğlu, O. (2013). Peynirlikönü Düdeni. *DELTA, 8,* 134–147.

Döker, M., Özakın, Y., Gürcan, G., & Erdoğu, B. (nd). *Kizilin Mağarası Arastirma Raporu*. Retrieved from np.

Döker, M., Özakın, Y., Öztekin, I., & Tuncer, G. (2004). EGMA Düdeni. *DELTA, 7,* 89–91.

Dumont, J., & Monod, O. (1976). Dipoyraz Masifinin Triyasik Karbonatlı Serisi. *MTA DERGİSİ, 87,* 26–41.

Eğrikavuk, C. (2006). *Çukurpınar Düdeni. DELTA, 7,* 84.

Eğrikavuk, M. (2013). *Keş Dağı Düdeni. OBRUK, 6,* 2–11.

Erdogan, S., Ergeneli, H., & Berberoglu, E. (2014). *Burdur Insuyu Mağarası*. Retrieved from Ankara.

Furman, A., & Özgül, A. (2004). The distribution of cave-dwelling bats and conservation status of underground habitats in Northwestern Turkey. *Biological Conservation, 120*(2), 243–248.

Gilli, E. (1979). Expedition Eynif 1979. *Spéléologie. Bull. Club Martel Nice, Club Alpin Français, 105,* 10–27.

Gilli, E. (1984). *Recherches sur le creusement et la stabilité des grands volumes karstiques souterrains. Applications au creusement es cavités artificielles de grande taille.* (PhD), University of Provence Aix-Marseille.

Gilli, E. (1986). Les grandes cavités souterraines: études et applications. *Karstologia, 7,* 2–10.

Gregory, A. (1989). *Taurus Mountains Expedition, August–September 1989*.

Gül, M. (2000). *Kahramanmaraş Yöresinin Jeolojisi.* (Ph.D.), Hacettepe Üniversitesi Fen Bilimleri Enstitüsü, Ankara.

Gül, M., Darbaş, G., & Gürbüz, K. (2005). Alacık Formasyonunun Kahramanmaraş Havzası İçindeki Tektono-stratigrafik Konumu. *Istanbul Üniv. Müh. Fak. Yerbilimleri Dergisi, 18*(2), 11–18.

Güldalı, N., & Nazik, L. (1984). Tınaz Tepe magara sistemi ve yakın çevresinin karstik gelişimi (The karstic development of the Tınaz Tepe cave system and its surrounding). *Jeomorfoloji Dergisi, 12,* 107–115.

Güldalı, N., Nazik, L., & Onal, O. (1980). *Akseki-Seydişehir Yörelerinin Önemli Mağaraları*. Retrieved from Ankara.

Güldalı, N., Nazik, L., Soylu, C., & Aksoy, B. (1987). *Alanya, Gazipaşa ve Manavgat Çevresinin Mağaraları*. Retrieved from Ankara.

Güldalı, N., Nazik, L., Soylu, C., Aksoy, B., & Beydeş, S. (1989). *Isparta-Beyşehir Yöresinde Mağara Araştırmaları*. Retrieved from Ankara.

Güldalı, N., Onal, O., & Nazik, L. (1982). *Tınaztepe Mağaraları Sistemi*. Retrieved from Ankara.

Güleç, E., Baykara, I., Özer, I., Sagir, M., Açıkkol, A., Erkman, A., & Yilmaz, H. (2010). *2008 Yılı Üçağızlı Mağarası Kazısı*. Paper presented at the 31. Kazı Sonuçları Toplantısı, Ankara.

Güleç, E., Kuhn, S., Ozer, I., & Steiner, M. (2003). *The 2001 Excavation Season at Üçağızlı Cave*. Paper presented at the 24th Kazı Sonuçları Toplantısı 2002, Ankara.

Güleç, E., Kuhn, S., Steiner, M., Ozer, I., Sagir, M., Erkman, A., ... Baykara, I. (2009). *2006–2007 Yılları Üçağızlı Mağarası Kazıları*. Paper presented at the 30th Kazı Sonuçları Toplantısı, Ankara.

Güngör, E., & Gülez, H. (1992). *Türkiye'nin Speleolojik Olanakları-1*. Retrieved from Ankara.

Gürcan, G., Yamaç, A., Kirlangiç, S., Pelen, M., Talay, Z., & Zorlu, P. (2006). *Türkiye Mağaralar Envanteri*. Istanbul: Ege Yayınları.

Halliday, W. (2001). *3100 Years at Turkey's Iskender-i Birkilin Caves*. Paper presented at the The Middle-East Speleology Symposium, Lebanon.

Halliday, W., & Shaw, T. (1995). The Iskender-i Birkilin Caves in the 9th and 12th Centuries BC. *The NSS Bulletin Journal of Caves and Karst Studies, 57,* 108–110.

Hardcastle, R., & Schmitt, G. (1987). Turkey—The mystery of Altınbeşik Düdensuyu Mağarası. *NSS News, 45*(4), 72–75.

Harmansah, Ö. (2007). 'Source of the Tigris'. Event, place and performance in the Assyrian landscapes of the Early Iron Age. *Archaeological Dialogues, 14*(2), 179–204.

Holland, M. (1990). Quest for the Ilgarini Mağarası, Expedition Report. *Caves and Caving, 52,* 24–27.

Hoşgören, M. (1981). Ege Bölgesinde Bir Karstik Yöre. *İstanbul Yerbilimleri, 2*(3–4).

Hovasse, R. (1927). Yarımburgaz Mağarası (La Grotte de Yarım Bourgas). *Istanbul Darulfününü Fen Fakültesi Mecmuası, 5,* 395–422.

Jäger, O., & Janoušek, E. (2013). *Geospeleos in Altınbesik Cave. OBRUK, 7*(54), 60.

Janousek, E. (2018). *Altınbeşik* Praha.

Jones, H. (1929). *The geography of Strabo* (Vol. VI, Books 12-13-14). London.

Kansu, Ş. (1966). Marmara ve Trakya Bölgesinde Tarih Öncesi Araştırmaları. *Belleten, 30*(119), 491–492.

Karaman, E. (1994). Isparta-Burdur Arasının Jeolojisi ve Tektonik Özellikleri. *Türkiye Jeoloji Bülteni, 37*(2), 119–134.

Kaufmann, G. (1993). The caves of Sofular and Çayırköy, Blacksea Coast; Turkey. *International Caver, 6,* 25–29.

Kay, P. (1990). *LUSS-Turkey 1990, Nothern Turkey, Cide Region East of Zonguldak on the coast of the Black sea.*

Kaya, B., Simsek, M., & Akis, A. (2015). Altınbeşik Mağarası Milli Parkı'nın (Ibradı/Antalya) Fiziki Coğrafya Özellikleri ve Ekoturizm Potansiyeli. *Turkish Studies, 10*(2), 521–544.

Kincaid, T., & Jablonski, J. (1996). *Project Karst Dive 95 Technical Report.*

Klimchouk, A., Bayari, S., Nazik, L., & Törk, K. (2006). Glacial destruction of cave systems in high mountains, with a special reference to the Aladaglar massif, Central Taurus, Turkey. *Acta Carsologica, 35*(2), 111–121.

Klimchouk, A., Kasjan, Y., Samokhin, G., Nazik, L., Bayari, S., Törk, K., & Ozel, E. (2005). *Kuzgun Cave: The first super-deep Cave in the Aladaglar Massif, Turkey.* Paper presented at the 14th International Congress of Speleology, Kalamos.

Klimchouk, A., Nazik, L., Bayari, S., Törk, K., & Kasjan, Y. (2016). Kuzgun Cave and its context: The first super-deep cave in the Aladaglar Massif, Turkey.

Kocacan, H. (1918). Bir Mağara Nasıl Tetkik Olunur ve Yarımburgaz Mağarasını Ziyaret. *Tedrisat Mecmuası, 61,* 12–18.

Kökten, I. (1955). Ein allgemeiner Überblick über die Prähistorischen Forschungen in Karain-Höhle bei Antalya. *Belleten, 75*(19), 284–293.

Kökten, I. (1958). Tarsus-Antalya Arası Sahil Şeridi Üzerinde ve Antalya Bölgesinde Yapılan Tarih Öncesi Araştırmaları. *Türk Arkeoloji Dergisi, 8*(2), 10–16.

Kökten, I. (1962). Maraş ve Antalya Vilayetinde Süreli Dip Tarih Araştırmaları Hakkında Kısa Bir Rapor. *Türk Arkeoloji Dergisi, 11* (1), 40–41.

Kökten, I. (1963). Istanbul Batısında Eskitaş (Paleolitik) Devrine Ait Yeni Buluntular. *Dil Tarih ve Coğrafya Fakültesi Dergisi, 20*(3–4), 277–278.

Kusch, H. (1993a). Die Tigrishöhlen in Ostanatolien (Türkei), Teil 1: Lage und Raumbeschreibungen. *Die Höhle, Zeitschrift für Karst und Höhlenkunde, 44*(3), 69–81.

Kusch, H. (1993b). Die Tigrishöhlen in Ostanatolien (Türkei), Teil 2: Erforschungsgeschichte und Archäologie. *Die Höhle, Zeitschrift für Karst und Höhlenkunde, 44*(4), 101–110.

Laumanns, M., & Özbek, O. (1990). Turkish—German Speleological activities in the area of Zonguldak 1988–1989. *MAD Bulletin, 7,* 25–29.

Lehmann-Haupt, C. (1901). Der Tigris-Tunnel. *Verhandlungen der Berliner Gesellschaft für Anthropologie, Ethnologie und Urgeschichte, 33,* 226–244.

Lehmann-Haupt, C. (1910). *Armenien Einst und Jetzt.* Leipzig.

Lindberg, K. (1952a). Notes sur Quelques Grottes de la Turquie avec Liste des Cyclopides (crustacés copépodes) recueillis dans ces grottes. *Annales de Spéléologie, VII*(1), 55–66.

Lindberg, K. (1952b). Un nouveau cyclopide (Crustace copépode) cavernicole de la Turquie d' Europe. *Bulletin de la Société zoologique de France, LXXVII,* 437-440.

Masriera, A. (1971). Erfolge der Spanischen Türkei—Expedition 1970. *Die Höhle, Zeitschrift für Karst und Höhlenkunde, 22*(1), 75–76.

Masriera, A., & Martorell, J. (1971). Memoria de la expedición espeleológica Turquía 1970 del Club Montañes Barcelonés. *Karst, 8*(29), 15–25.

Moretti, G. (1923–1924a). In Daghindà Quogia In, La Grande Caverna nella Montagna delle Caverne. *Annuario della Reale Scuola Archeologica di Atene, 6-7,* 509–546.

Moretti, G. (1923–1924b). Le grotte sacre di luvadja. *Annuario della Reale Scuola Archeologica di Atene, 6-7,* 547–554.

MTA (Cartographer). (1964). Geological Map of Zonguldak 1/500,000.

MTA (Cartographer). (2002a). Geological Map of Istanbul 1/500,000.

MTA (Cartographer). (2002b). Geological Map of Kayseri 1/500,000.

Müller, T. (1975). Türkei - Expedition 74. *Der Schlaz, Verein für Höhlenkunde, 14,* 4–6.

Nazik, L. (2004). *The karst regions of Turkey (according to morphogenesis and morphometric properties).* Paper presented at the International Symposium on Earth System Sciences, İstanbul.

Nazik, L., Derici, S., & Kutluay, H. (1999). *İnsuyu Mağarası Mimari Elektrifikasyon Uygulama Projesi.* Retrieved from Ankara.

Nazik, L., Derici, S., Kutluay, H., Törk, K., Ozel, E., & Tuncer, K. (1998). *Dodurgalar Mağarası (Dodurgalar-Acıpayam-Denizli) Araştırma Raporu.* Retrieved from Ankara.

Nazik, L., Güldali, N., Tüfekçi, K., Beydeş, S., & Aksoy, B. (1993). *Beyşehir ve Derebucak İlçelerinin (Konya) Doğal Mağaraları.* Retrieved from Ankara.

Nazik, L., Mengi, H., Özel, E., Bircan, A., & Beydeş, S. (1995). *Zonguldak Yakın Çevresinin Doğal Mağaraları.* Retrieved from Ankara.

Nazik, L., & Mulazimoglu, N. (2007). *Inventory and Possible categories of Caves in Turkey.* Paper presented at the 12th Regional Conference on Geoconservation and ProGEO Working Group Ljubljana, Slovenia.

Nazik, L., Poyraz, M., & Karabıyıkoğlu, M. (2019). Karstic landscapes and landforms in Turkey. In C. Kuzucuoğlu, A. Çiner, & N. Kazancı (Eds.), *Landscapes and landforms of Turkey* (pp. 181–196). Springer.

Nazik, L., & Törk, K. (2000). *Orta Torosların Karstik Gelişim Özellikleri.* Paper presented at the 53rd Türkiye Jeoloji Kurultayı Ankara.

Nazik, L., Törk, K., & Özel, E. (1997). *Güney Marmara Bölgesinin (Balıkesir; Bursa; Bilecik) Doğal Mağaraları.* Retrieved from Ankara.

Nazik, L., & Tuncer, K. (2010). Türkiye Karst Morfolojisinin Bölgesel Özellikleri. *Türk Speloloji Dergisi, Karst ve Mağara Araştırmaları, 1,* 6–15.

Öcal, T. (2012). Zindan Mağarasının Doğal Mirası ve Turizm Potansiyeli. *EKEV Akademi Dergisi, 52,* 279–298.

Özbek, M., & Güloğlu, M. (2005). A new cave Amphipod from Turkey: *Gammarus ustaoglui* sp. *Israel Journal of Zoology, 51,* 147–155.

Özdoğan, M. (1988). *Yarımburgaz Mağarası 1986 Yılı Kazı Çalışmaları*. Paper presented at the 5th Araştırma Sonuçları Toplantısı 1987, Ankara.

Özgül, A., & Bilgin, R. (2000). *Eurasian Bridge '99 - Bat Research in Northwestern Turkey*. Retrieved from Istanbul.

Öztürk, H. (2015). The re-evaluation of the Kocain (Antalya) Eirenarches, Anteirenarches and diogmites inscriptions. *Adalya, Akdeniz Medeniyetleri Araştırma Enstitüsü Dergisi, XVIII*, 159–180.

Öztürk, H. (2018). New inscriptions of Eirenarches and diogmites from Kocain *A. Vedat Çelgin'in 68. Doğum Günü Onuruna Makaleler* (pp. 741–744). Istanbul: Arkeoloji ve Sanat Yayınları.

Paksuz, S., Özkan, B., & Postawa, T. (2007). Seasonal Changes of Cave-dwelling bat fauna, and their relationship with microclimate in Dupnisa Cave System (Turkish Thrace). *Acta Zoologica Cracoviensia, 50A*(1–2), 57–66.

Raimond, M., & Chabert, C. (1971). Güvercin Tasi Deliği. *Grottes et Gouffres, Bulletin Périodique du Spéléo-Club de Paris, 48*, 36–37.

Russel, H. (1986). *Assyrian Monuments at the Tigris Tunnel*. Paper presented at the 3rd Araştırma Sonuçları Toplantısı, Ankara.

Samokhin, G. (2013). Exploration of Kuzgun Cave. *OBRUK, 7*, 30–35.

Schachner, A. (2005). *Birkleyn Mağaraları (Dicle Tüneli) Yüzey Araştırması*. Paper presented at the 23rd Araştırma Sonuçları Toplantısı, Antalya

Schachner, A. (2009). *Assyriens Könige an einer der Quellen des Tigris: archäologische Forschungen im Höhlensystem von Birkleyn und am sogenannten Tigris-Tunnel*. Tübingen Wasmuth Ernst Verlag.

Schmitt, G. (1976). Türkei expedition 1976. *Der Schlaz, Verein für Höhlenkunde, 20*, 11–16.

Schmitt, G. (1980). Caving in Turkey. *Caving International Magazine, 9*, 8–12.

Siler, M. (2016). *Anamur Çevresinin Karst Jeomorfolojisi*. (Ph.D.), Fırat University.

Spitzenberger, F. (1973). Höhlen in Westanatolien (Türkei). *Die Höhle, Zeitschrift für Karst und Höhlenkunde, 24*, 23–30.

Sroubek, P. (1989). *Taurus 89 Geospeleos Expedition Report*. Retrieved from Praha.

Stratford, T. (1991). Balatini Magarasi: Toros daglari, S-W Turkey. *International Caver, 1*, 16–22.

Stratford, T., & Crabbe, M. (1992). *Toros 89-92, The Çamlık Project*. Retrieved from Swindon.

Susam, S., & Genc, B. (1996). Çukurpınar Düdeni (Olucak; Anamur; Mersin; Turquie). *Spelunca, 63*, 19–20.

Şimşek, M. (2018). *Geyik Dağı Kütlesinde (Orta Toroslar) Karstik Depresyonların Dağılışı*. (Ph.D.), Istanbul University, Istanbul.

Taşkıran, H. (2011). Suluin Mağarası Kazıları 2007–2010 / Excavations at Suluin Cave in 2007–2010. *ANMED, 9*, 119–127.

Taşkıran, H., Kösem, B., Özçelik, K., Aydın, Y., & Erbil, E. (2015). *2013 Yılı Suluin Mağarası Kazısı*. Paper presented at the 36th Kazı Sonuçları Toplantısı, Ankara.

Taşkıran, H., Özçelik, K., Kösem, B., Aydın, Y., Yaman, I., & Karakoç, M. (2014). *2012 Yılı Suluin Mağarası Kazısı*. Paper presented at the 35h Kazı Sonuçları Toplantısı, Muğla.

Taylor, J. (1865). Travels in Kurdistan, with notices of the sources of the Eastern and Western Tigris, and ancient ruins in their neighbourhood. *Journal of the Royal Geographical Society of London, 35*, 21–58.

Tekeli, O., Aksay, A., Urgun, B., & Isik, A. (1983). *Geology of the Aladag Mountain*. Paper presented at the Geology of the Taurus Belt, Ankara.

Timing, H. (1984). Die Höhle in der Bildenden Kunst. *Die Höhle, Zeitschrift für Karst und Höhlenkunde, 35*(3/4), 177–190.

Törk, K. (1994). *Recession Curve Analysis of Hızar and Mencilis Karst Springs*. Paper presented at the 2nd Speleoloji Sempozyumu.

Törk, K. (1995). *Karst Hydrogeological Investigation of Northern Safranbolu*. (MD), Hacettepe Üniversitesi, Ankara.

Törk, K., Atabey, E., H, M., E, Ö., Aksoy, B., & Acar, C. (2001). *Çok dönemli-çok kökenli gelişimi karakterize eden mağaralara bir örnek: Oylat mağarası (İnegöl-Bursa)*. Paper presented at the 54th Türkiye Jeoloji Kurultayı, Ankara.

Törk, K., Ozel, E., Savas, F., Akgoz, M., Acar, C., & Tuncer, K. (2007). *Damlataş Mağarası (Saimbeyli-Adana) Araştırma Raporu*. Retrieved from Ankara.

Törk, K., & Savas, F. (2008). *Kahramanmaraş Mağaraları Araştırma Raporu*. Retrieved from Ankara.

Törk, K., Savas, F., Özel, E., Akgoz, M., Acar, C., Akçakaya, U., & Kahraman, İ. (2009). *Sıtmasuyu Mağarası Araştırma Raporu*. Retrieved from Ankara.

Törk, K., Savas, F., Yeleser, L., & Kahraman, İ. (2014). *Bayındır Mağarası (Bursa-Büyükorhan) Araştırma Raporu*. Retrieved from Ankara.

Tunoğlu, C. (1991). *Devrekani Kuzey Yöresinin Jeolojik İncelemesi*. (Ph.D.), Hacettepe Üniversitesi, Ankara.

Tutkun, S. (1989). Saimbeyli (Adana) Yöresinin Tektonik Özellikleri *Jeoloji Mühendisliği Dergisi, 34–35*, 57–63.

Usta, N., & Ipekoglu, H. (2019). An Anthropological assessment about burial practices of Kadiini Cave People (Alanya/Antalya). *Cedrus, 7*, 41–58.

Usuloğlu, E. (2019). *Morca 2019*.

Usuloğlu, E., & Siler, M. (2014). *Taşeli Projesi Kapsamında Çukurpınar Düdeni Araştırma Sonuçları*. Paper presented at the 7th Speleoloji Sempozyumu, Izmir.

Usuloğlu, E., & Siler, M. (2015). *Taşeli Platosu Çalışmaları 2011–2014*. Retrieved from Istanbul.

Waltham, A. (1976). The Tigris Tunnel and Birkleyn Caves, Turkey. *British Cave Research Association Bulletin, 14*, 31–34.

Watkins, P. (1980). *Trent Polytechnic Speleological Expedition to Turkey 1978*. Retrieved from Nottingham.

Wolozan, D. (1992). 4ème et 5ème expé de l'A.D.E.K.S. *Scialet, 21*, 184–194.

Wolozan, D. (1993a). Cinquième expédition spéléologique de l'A.D.E. K.S. en Turquie *Spelunca, 49*, 27–29.

Wolozan, D. (1993b). *Etudes et Documents de l'Adeks, Recherches Spéléologiques en Turquie 1992*. Retrieved from Lyon.

Wolozan, D. (1993c). *Recherches Spéléologiques en Turquie (4ème et 5ème campagnes de l'ADEKS)*.

Yalçınkaya, I. (1987). *1985 Yılı Karain Kazıları*. Paper presented at the 8. Kazı Sonuçları Toplantısı 1986, Ankara.

Yalçınkaya, I. (1992a). *1990 Yılı Karain Kazıları* Paper presented at the 13rd Kazı Sonuçları Toplantısı 1991, Ankara.

Yalçınkaya, I. (1992b). *1990 Yılı Öküzini Kazısı*. Paper presented at the 13rd Kazı Sonuçları Toplantısı 1991, Ankara.

Yalçınkaya, I., Leotard, J., Kartal, M., Erek, C., Atici, A., & Lopez, I. (1997). *1990–1995 Öküzini Kazıları (Tardiglasiyer Yerleşimleri)*. Paper presented at the 18. Kazı Sonuçları Toplantısı 1996, Ankara.

Yalçınkaya, I., Otte, M., Taşkıran, H., Kösem, B., & Ceylan, K. (1997). *1985–1995 Karain Kazıları Işığında Anadolu Paleolitiğinin Önemi*. Paper presented at the 18. Kazı Sonuçları Toplantısı 1996.

Yalçınkaya, I., Taşkıran, H., Kartal, M., Kösem, B., Ceylan, K., Erek, C., … Otte, M. (1998). *1995 ve 1996 Yılları Karain Kazıları*. Paper presented at the 19. Kazı Sonuçları Toplantısı 1997, Ankara.

Yalçınkaya, I., Taşkıran, H., Kösem, B., Özçelik, K., & Atıcı, A. (2002). *2000 Yılı Karain Kazısı*. Paper presented at the 23rd Kazı Sonuçları Toplantısı 2001, Ankara.
Yamaç, A. (1990). *Kocain mağarasının keşfi ve araştırma tarihi*. Paper presented at the 1st Speleoloji Sempozyumu, Bogazici University, Istanbul.
Yamaç, A. (2009). Yazören Mağarası Araştırması. *OBRUK, 1*, 2–5.
Yamaç, A. (2010). Sarpunalınca Mağarası. *OBRUK, 3*, 48–54.
Yamaç, A. (2011). Kahramanmaraş Bölgesi. *OBRUK, 4*, 4–12.
Yamaç, A. (2013a). *Anatolian Caves in Strabo's Geographica*. Paper presented at the 16th International Congress of Speleology, Brno.
Yamaç, A. (2013b). *Claude Chabert and the mapping of Ayvaini Cave—Turkey*. Paper presented at the 16th International Congress of Speleology, Brno
Yamaç, A. (2013c). Kocain Mağarası'nın Haritalanması. *OBRUK, 7*, 44–49.
Yamaç, A. (2013d). *Re-mapping of Insuyu Cave (Burdur—Western Turkey)*. Paper presented at the 16th International Congress of Speleology, Brno.
Yamaç, A., & Eğrikavuk, M. (2010). Küre Milli Parkı Araştırmaları. *OBRUK, 2*, 52–70.
Yamaç, A., & Eğrikavuk, M. (2013). Survey efforts in Insuyu Cave. *OBRUK, 7*, 6–25.
Yamaç, A., & Yıldız, E. (2011). *Exploration of Sarpunalınca Cave*. Paper presented at the 3rd Middle East Speleological Symposium, Beirut.
Yapici, N., Anil, M., & Yetis, C. (2003). Anamur-Malaklar (Mersin) Yöresinin Jeolojisi. *Çukurova Üniversitesi Mühendislik Mimarlık Fakültesi Dergisi, 18*(2), 267–282.
Yasar, Z., Demirel, F., & Cankaya, A. (2012). Burdur/ Kızılin Mağarası Iskeletlerinin Antropolojik Açıdan Değerlendirilmesi. *Çukurova Üniversitesi Sosyal Bilimler Enstitüsü Dergisi, 21*(2), 129–142.
Yümün, Z., & Kiliç, A. (2002). Kamandağı ile Camdere Köyü Arasının Stratigrafisi. *Cumhuriyet Üniversitesi Mühendislik Fakültesi Dergisi, Seri A-Yerbilimleri, 19*(2), 16–24.
Zeinelabdin, S. (2002). *An Evaluation of Conservation Strategies in Turkey with Particular Focus on Recent Developments in the Dupnisa Cave System.* (MA Thesis), Boğaziçi Üniversitesi, Istanbul.

Index

A. Yamaç et al., *Caves and Karst of Turkey—Vol. 1*, Cave and Karst Systems of the World,
https://doi.org/10.1007/978-3-030-65501-3